Graph Neural Network Methods and Applications in Scene Understanding

Weibin Liu · Huaqing Hao · Hui Wang ·
Zhiyuan Zou · Weiwei Xing

Graph Neural Network Methods and Applications in Scene Understanding

Springer

Weibin Liu
School of Computer Science
and Technology
Institute of Information Science
Beijing Jiaotong University
Beijing, China

Hui Wang
College of Electronic Information
Engineering
Hebei University
Hebei, China

Weiwei Xing
School of Software Engineering
Beijing Jiaotong University
Beijing, China

Huaqing Hao
College of Electronic Information
Engineering
Hebei University
Hebei, China

Zhiyuan Zou
Computer School
Beijing Information Science
and Technology University
Beijing, China

ISBN 978-981-97-9932-9 ISBN 978-981-97-9933-6 (eBook)
https://doi.org/10.1007/978-981-97-9933-6

This Springer imprint is published by the registered company Springer Nature Singapore Pte Ltd.
The registered company address is: 152 Beach Road, #21-01/04 Gateway East, Singapore 189721, Singapore

Preface

This book mainly introduces the graph neural network and its applications in scene understanding. Graph neural network is developed for feature learning of graph data, which is the most prevalent form of data in the real world. This book aims to introduce the main existing techniques of graph neural network and explore some innovative applications of graph neural networks to scene understanding. It can offer a comprehensive guide to leveraging this powerful combination for advancing computer vision and beyond. This book is designed to serve as a comprehensive guide, explaining in detail how graph neural networks can be applied to specific tasks in scene understanding and effectively solve the difficulties and challenges faced by the tasks. In general, this book is organized as follows:

Chapter 1 introduces the history and frontiers of graph neural networks. Chapter 2 provides a detailed introduction to the task of scene understanding and highlights the three key tasks that this book focuses on. Chapter 3 explains the basic knowledge of graph neural network and the classical architectures of graph neural network in detail. Chapter 4 mainly introduces the graph convolutional network for scene parsing. Chapter 5 explores how graph neural networks can be applied to both strongly supervised and weakly supervised human parsing. Chapter 6 introduces the dynamic graph construction for human parsing and the dynamic graph pooling for human parsing. Chapter 7 introduces the graph neural networks for video object segmentation. Chapter 8 summarizes the book and proposes some future research directions for scene understanding tasks.

Beijing, China Weibin Liu
June 2024

Acknowledgements

First of all, we would like to thank everyone of the collaborators who work tirelessly together and contribute to producing this monograph. We would also like to thank the organizers of scene understanding benchmarks who provide us the large-scale datasets and benchmark for comprehensive evaluating and assessing the different types of scene understanding algorithms. We would highly appreciate the help and assistance of the current and graduated students in our research group during the production of this book. In particular, we wish to thank Mr. Jicheng Zhu and Mr. Shaobin Fu for their time and efforts in providing materials and reviewing for revising the entire book and improving the individual chapters. We also want to especially thank the editor of Springer, Jing Dou, for her patience, support, guidance and editorial assistance in the course of the preparation of this work. Part of the content in this book is based on the work supported by the National Natural Science Foundation of China under No. 61876018, and the Beijing Natural Science Foundation under No. 4212025 and No. L231005.

This monograph, especially those proposed scene parsing algorithms, human parsing algorithms and video object segmentation algorithms in Chaps. 4 to 7, are mainly prepared based on the following doctoral dissertations of our group, we would like to give a special acknowledgement to all the authors for their contributions. Besides, this monograph is also based on the following research papers from our group, we would like to give a special acknowledgement to all the authors for their contributions and to the related publishing organizations for their permission to use these materials. The specific referenced papers are shown below:

1. Hao Huaqing. Research on Human Parsing Algorithm based on Graph Neural Network [D]. Beijing Jiaotong University, 2024.
2. Wang Hui. Research on Video Object Segmentation Methods under Complex Dynamic Scenes [D]. Beijing Jiaotong University, 2023.
3. Zou Zhiyuan. Research on Feature Fusion Methods for Scene Recognition and Understanding [D]. Beijing Jiaotong University, 2022.

4. Huaqing Hao, Weibin Liu[*], Weiwei Xing, et al. Multilabel Learning based Adaptive Graph Convolutional Network for Human Parsing [J]. Pattern Recognition, 2022, 127:108593.
5. Huaqing Hao, Weibin Liu[*], Weiwei Xing. Context Prior based Semantic-spatial Graph Network for Human Parsing [J]. Neurocomputing, 2021, 457:13–25.

Contents

About the Authors

Weibin Liu received the Ph.D. degree in Signal and Information Processing from Institute of Information Science at Beijing Jiaotong University, China, in 2001. During 2001–2005, he was a researcher in Information Technology Division at Fujitsu Research and Development Center Co., Ltd. Since 2005, he has been with the Institute of Information Science, School of Computer Science and Technology at Beijing Jiaotong University, where currently he is a professor in Digital Media Research Group. He was also a visiting researcher in Center for Human Modeling and Simulation at University of Pennsylvania, PA, USA during 2009–2010. His research interests include computer vision, video and image processing, deep learning, computer graphics, virtual human and virtual environment, and pattern recognition.

Huaqing Hao received the B.S. and M.S. degree in Electronic Information Engineering from Heibei University, China, in 2015 and 2018, respectively. She received the Ph.D. degree in Signal and Information Processing from Institute of Information Science at Beijing Jiaotong University, China, in 2024. Currently, she is an associate professor at College of Electronic Information Engineering, Hebei University. Her main research interests include computer vision, pattern recognition and deep learning, in particular focusing on human parsing.

Hui Wang received the B.S. degree in Electronic Information Engineering from Hebei University, China, in 2016. He received the Ph.D. degree in Signal and Information Processing from Institute of Information Science at Beijing Jiaotong University, China, in 2023. Currently, he is an associate professor at College of Electronic Information Engineering, Hebei University. His research interests include computer vision, image processing, video object segmentation.

Zhiyuan Zou received the B.S. degree in Software Engineering from Beijing Jiaotong University, Beijing, China, in 2015, and Ph.D. degree in Software Engineering from Institute of Information Science, Beijing Jiaotong University, in

2022. Currently, he is an associate professor at Computer School, Beijing Information Science and Technology University. His research interests include scene understanding, deep learning, computer vision, and pattern recognition.

Weiwei Xing received the B.S. degree in Computer Science and Technology and the Ph.D. degree in Signal and Information Processing from Beijing Jiaotong University, Beijing, China, in 2001 and 2006, respectively. She was a visiting scholar at University of Pennsylvania, PA, USA during 2011–2012. She is currently a professor at School of Software Engineering, Beijing Jiaotong University and leads the research group on Intelligent Computing and Big Data. Her research interests include computer vision, video and image processing, deep learning, and big data analysis.

Chapter 1
Introduction

In the modern era, machine learning has experienced a revolutionary transformation, driven by the rise of graph representation learning and the subsequent evolution of graph neural networks (GNNs). Graph data from different domains, from social networks and molecular structures to recommendation systems and knowledge graphs, have catalyzed this shift. Conventional machine learning methods frequently fall short in capturing the intricate relationships and interdependencies present within these graph data.

Graphs, which are versatile mathematical structures made up of nodes and edges, provide a powerful framework for depicting relationships and associations within graph data. Yet, extracting the information embedded within these graphs in an efficient manner presents a notable hurdle. This very challenge has spurred the evolution of graph representation learning—a sub-field within machine learning devoted to acquiring expressive and informative node and edge representations from graphs.

Graph neural networks have greatly contributed to the development of graph representation learning. These specialized architectures, building upon the core principles of neural networks, have progressively adapted to the intricacies inherent in data structured as graphs. Graph neural networks have showcased remarkable capabilities across various undertakings, including node classification, link prediction, and graph classification. These approaches involve a cyclic process of gathering information from neighboring nodes, a mechanism that captures both local and global patterns. The information is then encoded into distributed representations, which, in turn, empowers the execution of subsequent machine learning tasks with remarkable effectiveness.

This chapter first introduces graph representation learning (Sect. 1.1), including an overview of graphs, graph signal processing, and graph representation methods.

W. Liu et al., *Graph Neural Network Methods and Applications in Scene Understanding*,
https://doi.org/10.1007/978-981-97-9933-6_1

We then introduce graph neural networks in Sect. 1.2, including their origin, development and current state-of-the-art approaches. We provide an outline of this book in Sect. 1.3 for easy reading.

1.1 Graph Representation Learning

Graph theory is a branch of mathematics whose main research object is graphs. A graph is made up of vertices and edges to describe a particular relationship between things, where different vertices represent different things, and connecting two vertices with an edge indicates that the two vertices have this relationship. Graph theory mainly studies various properties (such as structural properties, algebraic properties, etc.) of graphs and their generalized forms (such as directed graphs, hyper-graph, colored graphs, etc.), as well as applications of graph theory in other disciplines or in real life. Graph theory is closely related to combination, number theory, algebra, analysis, geometry, topology, operations research, probability theory and other mathematical disciplines, and has a wide range of applications in physics, chemistry, biology, computer science, logic and other disciplines.

A graph is a data structure with a wide range of meanings, which is used to represent relationships between objects. In mathematics, graphs are the main object of study in graph theory. In computer engineering, graphs are a common data structure. There are very many real-world systems in life that correspond to graphs, such as chemical molecules, communication networks, social networks, and knowledge graphs. Any system that contains binary relations can be described by a graph. On the other hand, a large number of real-world problems can be transformed into computational tasks to be solved on graphs. Many theories of graph learning focus on tasks related to graph data.

The main goal of graph representation learning is to transform graph data into a low dimensional dense vectorization representation, and to ensure that the properties of graph data can be matched in vector space. The representation of graph data can be at the node level or the whole graph level. The representation learning of nodes (the basic elements of graph data) has always been the main object of graph representation learning. If the representation of graph data contains rich semantic information, good node features can be obtained, and linear classifiers can be directly selected to learn the classification task.

In simple terms, graph representation learning is to map nodes to vector representations and retain as much topological information as possible. Graph representation learning is mainly divided into representation learning based on graph structure and representation learning based on graph features. Among them, the vector representation of nodes in graph structure-based representation learning only comes from the topological structure of the graph, and lacks the representation of the feature messages of graph nodes. The vector representation of nodes based on graph feature learning contains both the topological information of graphs and the existing feature vectors.

In graph representation learning, vector spaces are also adjacent when they are adjacent on the graph. The second proximity is the distance between two vectors in the Euclidean space. There are several explanations for the first proximity as follows: 1-hop (first-order neighbors), two adjacent nodes can be defined as near; K-hop (K-order neighbors), two adjacent nodes of order k can also be defined as adjacent. For example, DeepWalk uses random walk for node sampling. In the social network application scenario, the graph structure can be constructed according to the user's behavior, and then the node sequence can be constructed by random sampling through random walk (for example, starting at point A, A-B, B then jumping to neighbor point E, and finally to F to get A-B-E-F sequence). The sequence problem is the language model in NLP, where a sentence is a sequence of words. After that, the problem of Word2vec (word represented by vector) can be obtained by Skip-gram model. This translates the graph structure into a sequence problem. Note that the probability of a node going to its neighbor is equal. Wandering patterns are different in directed and undirected graphs. The walking mode in undirected graph is connected to walk; In a directed graph, you only go in the direction of the outgoing edge.

1.1.1 Basic Concepts of Graphs

A graph can be represented as $G = (V, E)$, where V denotes the set of nodes and E denotes the set of edges connecting these nodes. We represent an edge that connects a node $q \in V$ to a node $v \in V$ as $(q, v) \in E$. Frequently, our focus will be primarily on simple graphs, where, at most, a single edge exists between any pair of nodes. Such graphs also exhibit absence of edges between a node and itself, and all edges are undirected, meaning if q is connected to v, then $(q, v) \in E \leftrightarrow (v, q) \in E$.

An adjacency matrix $\mathbf{A} \in \mathbb{R}^{|V|\times|V|}$ provides a convenient means of graph representation. We arrange the nodes in order, ensuring each node corresponds to a specific row and column within the matrix. This arrangement facilitates the depiction of edges as matrix entries: $\mathbf{A}[q, v] = 1$ if there exists an edge $(q, v) \in E$, while $\mathbf{A}[q, v] = 0$ if no such edge exists. $\mathbf{A}[q, v]$ can also be written as $\mathbf{A}_{q,v}$. If the graph comprises solely of undirected edges, matrix A will exhibit symmetry. However, for a directed graph, matrix A might not maintain its symmetry. Additionally, graphs can incorporate weighted edges, leading to entries in the adjacency matrix taking on real-values rather than being confined to binary values (0 or 1).

For a graph $G = (V, E)$, the degree of node v_i is shown in Eq. (1.1), which indicates the number of nodes connected to node v_i.

$$d(v_i) = \sum_{v_j \in V} \mathbb{I}_E\left(\{v_i, v_j\}\right), \tag{1.1}$$

where $\mathbb{I}_E\left(\{v_i, v_j\}\right) = 1$ if $(v_i, v_j) \in E$ and $\mathbb{I}_E\left(\{v_i, v_j\}\right) = 0$ otherwise. We can also use equation $d(v_i) = \sum_{j=1}^{N} \mathbf{A}_{i,j}$ to calculate the degree of node v_i.

A graph walk constitutes a series of nodes and edges taken in an alternating fashion, commencing with a node and concluding with a node. A q-v walk means that a walk originates from node q and ends at node v. It's important to note that q-v walks are not exclusive, as multiple variations of such walks with differing edge counts can exist. A path is a type of walk where the nodes traversed are unique and not repeated. A trail is a type of walk where the edges traversed are unique and not repeated.

A subgraph, denoted as $G_0 = (V_0, E_0)$, of a provided graph $G = (V, E)$, is created by selecting a subset of edges E_0 from E and a subset of nodes V_0 from V.

1.1.2 Graph Statistical Methods

The node degree provides a basic quantification of a node's connectivity by counting its neighbors. However, this metric alone may not adequately capture a node's significance within a graph. We can explore diverse measures of node centrality to assess a node's importance. Eigenvector centrality is one important centrality metric, which incorporates the importance of a node's neighbors as well. The eigenvector centrality e_c of a node v_i is determined by the average centrality of its neighboring nodes.

$$e_c\left(v_i\right) = \frac{1}{\lambda} \sum_{v_j \in \mathcal{N}(v_i)} \mathbf{A}_{i,j} e_c\left(v_j\right), \tag{1.2}$$

where λ represents a constant. When we rephrase this equation using vector notation and denote $\mathbf{e}$ as the vector containing node centralities, it becomes apparent that this recursive formulation corresponds to the conventional eigenvector equation applied to the adjacency matrix: $\lambda\mathbf{e} = \mathbf{A}\mathbf{e}$.

Evidently, $\mathbf{e}$ represents an eigenvector of matrix $\mathbf{A}$, associated with the eigenvalue λ. For an adjacency matrix $\mathbf{A}$, multiple eigenvalue-eigenvector pairs exist. Applying the Perron-Frobenius theorem, we can opt to designate λ as the largest eigenvalue, with the corresponding eigenvector serving as the vector representing centrality scores. An alternative approach to assessing a node's importance involves evaluating its position importance within the graph's structure. If a node is passed through by numerous paths traversing the graph, it holds an important position. This concept encompasses centrality metrics like betweenness centrality, which quantifies the frequency of a node's occurrence on the shortest path connecting two other nodes, and closeness centrality, which quantifies the average shortest path length between a node and all other nodes. Newman [1] provides a comprehensive review of these measures and many others.

Up to this point, we have explored some representative metrics and attributes focusing on individual nodes. These could serve as valuable attributes for tasks involving the classification of nodes. Nevertheless, what if our objective shifts towards classifying entire graphs? We will provide a concise overview of some methodologies employed to extract features at the graph level for such tasks. A considerable portion of the techniques we explore in the following belong to the broader category of graph kernel methods. These methods encompass strategies for designing features for graphs or generating implicit kernel functions that are applicable in machine learning models. We put emphasis on techniques that derive clear and direct feature representations, rather than methods that establish implicit kernels between graphs.

Bag of nodes is the most straightforward method for conceptualizing a graph-level feature, which involves aggregating statistics from individual nodes. For instance, one could calculate histograms or other condensed statistics pertaining to factors like node degrees, centralities, and clustering coefficients within the graph. This amalgamated data can subsequently serve as a representation at the graph level. Nonetheless, a drawback of this strategy lies in its complete reliance on localized node-level data, missing global information of the graph.

An enhanced method for advancing the fundamental bag of nodes involves employing an iterative strategy for neighborhood aggregation. The concept behind these methodologies revolves around extracting node-level features that encompass more comprehensive information beyond their immediate local ego graph. Subsequently, these enriched features are aggregated into a comprehensive representation at graph-level. Among these strategies, the Weisfeiler-Lehman (WL) algorithm and kernel [2, 3] stand out as arguably the most pivotal and widely recognized.

The WL algorithm sequentially assigns novel labels to individual nodes by applying a hash function to the multi-set of existing labels found within the neighborhood of each node. This hash function is responsible for converting distinctive multi-sets into distinct new labels.

Upon completing M rounds of re-labeling, labels have been assigned to each node, encapsulating the information of its neighborhood within M-hops. Consequently, these labels can be utilized to generate feature representations for the graph, either through histograms or alternative summary statistics. Put simply, the Weisfeiler-Lehman (WL) kernel is determined by quantifying the disparity between the two resulting label sets.

Similar to our exploration of node-level attributes, a sound and potent approach to crafting graph-level attributes is to tally instances of various compact subgraph structures, often referred to as graphlets. In a precise sense, the graphlet kernel entails the exhaustive listing of all potential graph structures of a specific size, followed by a computation of their frequency within the full graph. The intricacy with this method lies in the fact that enumerating these graphlets entails a demanding combinatorial task, although several approximations have been put forth to address this challenge.

A substitute to the exhaustive listing of all potential graphlets is to adopt path-based techniques. In these methodologies, instead of cataloging graphlets, the focus is placed on investigating the assorted varieties of pathways present within the graph. To illustrate, the random walk kernel introduced by Kashima et al. [4] entails executing random walks across the graph and subsequently tabulating the instances of distinct degree sequences. Conversely, Borgwardt and Kriegel's shortest-path kernel

[5] follows a similar notion but confines its consideration to the shortest paths between nodes, excluding random walks. This concept of characterizing graphs through walks and paths wields considerable efficacy, enabling the extraction of structural features while circumventing numerous of the intricate combinatorial challenges inherent in graph data.

We have delved into several methodologies for extracting features or statistics from graph data at node- or graph-level. These obtained features prove advantageous across a spectrum of classification tasks. Nevertheless, their scope is restricted due to their inability to quantify inter-node relationships. If we need to accomplish the relationship prediction between nodes, the above features exhibit limited efficacy. Consider the most basic measure of neighborhood overlap, which merely tallies the common neighbors shared by two nodes:

$$\mathbf{O}[q, v] = |\mathcal{N}(q) \cap \mathcal{N}(v)|, \tag{1.3}$$

where we denote the quantification of the relationship between nodes q and v as $\mathbf{O}[q, v]$, the similarity matrix $\mathbf{O} \in \mathbb{R}^{|V| \times |V|}$ summarizes pairwise statistics of all the nodes.

Local overlap statistics are usually calculated based on the common neighbors shared by two nodes. The entries of the Sorensen index $\mathbf{O}_{\text{sor}} \in \mathbb{R}^{|V| \times |V|}$ representing node to node neighborhood overlaps are obtained by Eq. (1.4).

$$\mathbf{O}_{\text{sor}}[q, v] = \frac{2|\mathcal{N}(q) \cap \mathcal{N}(v)|}{d_q + d_v}. \tag{1.4}$$

The number of common neighbors is normalized by the sum of the degrees of nodes q and v. Normalization plays a crucial role because it prevents the overlap measure from being biased towards nodes with high degrees. Similarly, the Salton index normalizes by the square root of the product of d_q and d_v, represented as:

$$\mathbf{O}_{\text{sal}}[q, v] = \frac{2|\mathcal{N}(q) \cap \mathcal{N}(v)|}{\sqrt{d_q d_v}}. \tag{1.5}$$

Additionally, there is the Jaccard overlap, computed as:

$$\mathbf{O}_{\text{jac}}[q, v] = \frac{|\mathcal{N}(q) \cap \mathcal{N}(v)|}{|\mathcal{N}(q) \cup \mathcal{N}(v)|}. \tag{1.6}$$

These various measures aim to capture and quantify the overlap between node neighborhoods while simultaneously mitigating distortions introduced by differences in node degrees. The literature [6] offers numerous variations of this approach. Some metrics extend beyond mere tallying of common neighbors, aiming to incorporate the significance or relevance of shared neighbors in their evaluation, such as the Resource Allocation index and the Adamic-Adar index.

Local overlap metrics achieve competitive performance even pitted against sophisticated deep learning techniques. Nevertheless, these local methods possess a constraint in that they solely focus on assessing local neighborhoods. To illustrate, two nodes might exhibit no local overlap within node neighborhoods, yet they could belong to the same community within the larger graph. Global overlap statistics consider these broader relationships to address this limitation.

The Katz index serves as a fundamental global overlap statistic. Its calculation involves a straightforward process of tallying the number of paths between a pair of nodes across all possible lengths. This computation is expressed through the following equation:

$$\mathbf{O}_{\text{katz}}[q, v] = \sum_{i=1}^{\infty} \beta^i \mathbf{A}^i[q, v]. \tag{1.7}$$

Here, i denotes the length of a path from node q to node v. The parameter $\beta \in \mathbb{R}^+$ is user-defined and plays a pivotal role in determining the emphasis placed on shorter versus longer paths. When β is set to a small value ($\beta < 1$), it diminishes the significance of longer paths. It's noteworthy that the Katz index belongs to the category of geometric series of matrices, a type of mathematical construct frequently encountered in the realms of graph representation learning.

The matrix geometric solution is determined by the theorem: Consider a square, real-valued matrix denoted as $\mathbf{X}$, where λ_e represents its largest eigenvalue. The solution is expressed as follows:

$$(\mathbf{I} - \mathbf{X})^{-1} = \sum_{i=0}^{\infty} \mathbf{X}^i. \tag{1.8}$$

This equation holds true if and only if $(\mathbf{I} - \mathbf{X})$ is non-singular and $\lambda_e < 1$. Building upon this insight, we can determine the solution to the Katz index as follows:

$$\mathbf{O}_{\text{katz}} = (\mathbf{I} - \beta\mathbf{A})^{-1} - \mathbf{I}. \tag{1.9}$$

In this equation, $\mathbf{O}_{\text{katz}} \in \mathbb{R}^{|V| \times |V|}$ represents the node-node similarity matrix.

1.1.3 Graph Spectral Methods

Let's delve into the challenge of learning to cluster nodes within a graph, a task that serves as a driving force behind the pursuit of acquiring low-dimensional node embeddings. We will start by introducing key matrices employed for graph representation and providing a concise primer on the spectral graph theory.

While adjacency matrices faithfully represent original graphs, there exist other matrix representations known as Laplacians, which possess valuable properties. These Laplacian matrices are obtained from the adjacency matrix by different transformations. Among these, the unnormalized Laplacian is calculated using the following equation:

$$\mathbf{L} = \mathbf{D} - \mathbf{A}, \tag{1.10}$$

where $\mathbf{D}$ and $\mathbf{A}$ represent the degree matrix and the adjacency matrix, respectively. For a simple graph, the Laplacian matrix exhibits several critical properties. It possesses symmetry and positive semi-definiteness. Furthermore, it encompasses $|V|$ non-negative eigenvalues, denoted as $0 = \lambda_{|V|} \leq \lambda_{|V|-1} \leq \cdots \leq \lambda_1$. The normalized variants of the Laplacian matrix share similar properties with the standard Laplacian, but they exhibit slight variations in their algebraic characteristics due to the applied normalization. These Laplacian variants can be advantageous for various graph analysis tasks.

The eigenvectors associated with the eigenvalue of zero in the Laplacian can be employed to group nodes into clusters according to the connected components. However, this technique is limited to clustering nodes within disconnected components. We extend this concept further, for a fully connected graph, the optimal node clustering can still be achieved through Laplacian.

To lay the foundation for the spectral clustering method, it's essential to establish the concept of an optimal cluster. Consider a subset of nodes $\mathcal{A} \in V$ in the graph, with its complement denoted as $\bar{\mathcal{A}}$, meaning that $\mathcal{A} \cup \bar{\mathcal{A}} = V$ and $\mathcal{A} \cap \bar{\mathcal{A}} = \emptyset$. When the graph is partitioned into M non-overlapping subsets, denoted as $\mathcal{A}_1, \ldots, \mathcal{A}_K$, the cut value is calculated by

$$\mathrm{Cut}(\mathcal{A}_1, \ldots, \mathcal{A}_M) = \frac{1}{2} \sum_{m=1}^{M} |(q, v) \in E : q \in \mathcal{A}_m, v \in \bar{\mathcal{A}}_m|. \tag{1.11}$$

The "Cut" represents the number of edges that intersect the boundary between node partitions. Consequently, one approach to determining the best way to cluster nodes into M groups is to choose a partition that minimizes this cut count. The concept of Ratio Cut goes a step further by ensuring that these partitions are reasonably substantial. By using Laplacian eigenvalue decomposition, an optimal assignment of clusters according to the Ratio Cut criterion [7] can be obtained.

Nevertheless, the methods outlined above, particularly those relying on node or graph-level statistics, have their limitations because they rely on meticulously crafted, manual statistics and metrics. These manually engineered features are rigid in nature, incapable of adapting through a training process, and their design can be a laborious and costly endeavor.

The subsequent focus revolves around techniques aimed at acquiring node embeddings through learning. These methods aim to represent nodes in lower dimensions, encapsulating both their position and local structure. The aim is to map node features into a latent space where the spatial relationships correspond to the connections, such as edges, in the original graph.

1.1.4 Graph Embedding Methods

Graph representation learning generally involves two fundamental operations. First, node attributes are mapped by an extractor model into low-dimensional vectors. A reconstructor model is then employed to recover the local information of each node within the original graph from obtained node vectors.

In its simplest form, the extractor is defined as $\text{EXT} : V \rightarrow \mathbb{R}^d$, and the common approach involves defining pairwise reconstructors with the signature $\text{REC} : \mathbb{R}^d \times \mathbb{R}^d \rightarrow \mathbb{R}^+$. Both the extractor and reconstructor are optimized to minimize the loss function, expressed as:

$$\text{REC}(\text{EXT}(q), \text{EXT}(v)) = \text{REC}(\mathbf{z}_q, \mathbf{z}_v) \approx \mathbf{S}[q, v]. \tag{1.12}$$

In this context, $\mathbf{z}_q$ and $\mathbf{z}_v$ denote the features of nodes q and v, respectively. $\mathbf{S}[q, v]$ represents the similarity value between nodes, obtained based on graph statistics.

Matrix factorization is employed to acquire a low-dimensional approximate representation of a node-level similarity matrix, denoted as $\mathbf{S}$. This matrix $\mathbf{S}$ extends beyond the typical adjacency matrix and encompasses a user-defined concept of similarity between nodes.

Laplacian eigenmaps stand as one of the earliest and highly influential factorization-based methods, drawing upon the foundational principles of spectral clustering. In this approach, the reconstructor is defined using node-node L2-distance:

$$\text{REC}(\mathbf{z}_q, \mathbf{z}_v) = \|\mathbf{z}_q - \mathbf{z}_v\|_2^2. \tag{1.13}$$

The objective function $\mathcal{L}$ then combines training pairs $\mathcal{D}$ of nodes based on their graph similarity:

$$\mathcal{L} = \sum_{(q,v)\in\mathcal{D}} \text{REC}(\mathbf{z}_q, \mathbf{z}_v) \cdot \mathbf{S}[q, v]. \tag{1.14}$$

In more recent work following Laplacian eigenmaps, a reconstructor based on inner product is commonly employed: $\text{REC}(\mathbf{z}_q, \mathbf{z}_v) = \mathbf{z}_q^\top \mathbf{z}_v$. Several node embedding methods fall into this particular style, including the Graph Factorization method [8], HOPE [9], and GraRep [10]. Each of these three methods incorporates the inner-product reconstructor alongside the subsequent mean-squared error formulation:

$$\mathcal{L} = \sum_{(q,v)\in\mathcal{D}} \|\text{REC}(\mathbf{z}_q, \mathbf{z}_v) - \mathbf{S}[q, v]\|_2^2. \tag{1.15}$$

Their primary distinctions lie in how they define the term $\mathbf{S}[q, v]$. Specifically, the adjacency matrix or general neighborhood overlap measures are utilized to define $\mathbf{S}$.

Building on these successes, recent years have seen a surge in successful methods that adapt the inner-product approach to use stochastic measures of neighborhood overlap. The key innovation in these approaches is that node embeddings are optimized so that two nodes have similar embeddings if they tend to co-occur on short random walks over the graph.

Similar to the matrix factorization approaches described above, DeepWalk [11] and node2vec [12] use a shallow embedding approach and an inner-product decoder. The key distinction in these methods is in how they define the notions of node similarity and neighborhood reconstruction. Instead of directly reconstructing the adjacency matrix $\mathbf{A}$ or some deterministic function of $\mathbf{A}$, these approaches optimize embeddings to encode the statistics of random walks. Mathematically, the goal is to learn embeddings so that the following (roughly) holds:

$$\begin{aligned} \text{REC}(\mathbf{z}_q, \mathbf{z}_v) &\triangleq \frac{e^{\mathbf{z}_q^\top \mathbf{z}_v}}{\sum_{v_k \in V} e^{\mathbf{z}_q^\top \mathbf{z}_k}} \\ &\approx p_{\mathcal{G},T}(v|q), \end{aligned} \tag{1.16}$$

where $p_{\mathcal{G},T}(v|q)$ is the probability of visiting v on a length-T random walk starting at q, with T usually defined to be in the range $T \in \{2, \ldots, 10\}$. Again, a key difference between Eq. (1.16) and the factorization-based approaches is that the similarity measure in Eq. (1.16) is both stochastic and asymmetric. To train random walk embeddings, the general strategy is to use the decoder from Equation and minimize the following cross-entropy loss:

$$\mathcal{L} = \sum_{(q,v) \in \mathcal{D}} -\log(\text{REC}(\mathbf{z}_q, \mathbf{z}_v)). \tag{1.17}$$

Here, we use $\mathcal{D}$ to denote the training set of random walks, which is generated by sampling random walks starting from each node. For example, we can assume that N pairs of co-occurring nodes for each node u are sampled from the distribution $(q, v) \sim p_{\mathcal{G},T}(v|q)$. DeepWalk employs a hierarchical softmax to approximate Equation, which involves leveraging a binary-tree structure to accelerate the computation. On the other hand, node2vec employs a noise contrastive approach to approximate Equation.

The LINE [13] approach does not explicitly leverage random walks, but it shares conceptual motivations with DeepWalk and node2vec. The basic idea in LINE is to combine two encoder-decoder objectives. The first objective aims to encode first-order adjacency information. The second objective is more similar to the random walk approaches. It is the same decoder as Eq. (1.16), but it is trained using the KL-divergence to encode two-hop adjacency information.

It can be shown that random walk methods are actually closely related to matrix factorization approaches. This reveals that the embeddings learned by DeepWalk are in fact closely related to the spectral clustering embeddings. The key difference is that the DeepWalk embeddings control the influence of different eigenvalues through T, i.e., the length of the random walk. Qiu et al. [14] derive similar connections to matrix factorization for node2vec and discuss other related factorization-based approaches inspired by this connection.

We have described some shallow embedding methods. It is also important to note that shallow embedding approaches suffer from some important drawbacks: The first issue is that shallow embedding methods do not share any parameters between nodes in the encoder, since the encoder directly optimizes a unique embedding vector for each node. This lack of parameter sharing is both statistically and computationally inefficient. From a statistical perspective, parameter sharing can improve the efficiency of learning and also act as a powerful form of regularization. From the computational perspective, the lack of parameter sharing means that the number of parameters in shallow embedding methods necessarily grows as $O(|V|)$, which can be intractable in massive graphs.

A second key issue with shallow embedding approaches is that they cannot leverage node features in the encoder. Many graph datasets have rich feature information, which could potentially be informative in the encoding process. Lastly, and perhaps most importantly, shallow embedding methods are inherently transductive. These methods can only generate embeddings for nodes that were present during the training phase. Generating embeddings for new nodes, which are observed after the training phase, is not possible unless additional optimizations are performed to learn the embeddings for these nodes. This restriction prevents shallow embedding methods from being used on inductive applications, which involve generalizing to unseen nodes after training.

1.2 Graph Neural Networks

The rapid advancement of deep learning has led to substantial progress and success in Graph Neural Networks (GNNs). GNNs aim to design a class of neural network architectures for graph-structured data which is composed of graph nodes, graph edges and adjacency matrix. At present, GNNs mainly have two operation methods, i.e., Graph Convolutional Network and Graph Attention Network. GNNs have demonstrated strong performance in many tasks, and for different tasks, GNNs have developed into graph generation models and dynamic graph neural networks.

1.2.1 Graph Neural Network

Graph neural networks aim to expand upon conventional neural networks to handle graph-structured data. The initial GNN concept can be attributed to [15–17]. To avoid confusion, we use GNN† to denote the initial graph neural network. In the initial GNN† [17], the node embedding feature $\mathbf{h}_i^{(t)}$ and output feature $\mathbf{o}_i$ are obtained by the following equation:

$$\mathbf{h}_i^{(t)} = f(\mathbf{x}_i, \mathbf{x}^{\mathbf{e}}_{(i,j)}, \mathbf{h}_j^{(t-1)}, \mathbf{x}_j), \tag{1.18}$$

$$\mathbf{o}_i = g(\mathbf{h}_i, \mathbf{x}_v^{(T)}), \tag{1.19}$$

where t denotes the iteration index. f and g denote the learnable parameter functions and serve as transition and output functions, respectively. f is executed iteratively. As a contraction mapping, f can reduce the distance between node features to ensure convergence. f can be applied to all nodes, regardless of variations in the number of neighbors or the absence of neighborhood ordering information. $\mathbf{x}$ represents the input node feature. $j \in N(i)$ denotes the neighbor node of node i. $\mathbf{x}^{\mathbf{e}}_{(i,j)}$ denotes the feature of the connecting edge (i, j). $\mathbf{h}_i^{(t)}$ is also referred to as a hidden state. The initial state $\mathbf{h}_i^0$ is randomly set.

From fixed point theory [18], it is known that f is a compression mapping that allows the iterative process shown in Eq. (1.18) to converge after a certain number of steps. Once the convergence criterion is met, the hidden states of the last step's nodes are processed by the output layer, as shown in Eq. (1.19). To train the model parameters, the Jacobi method [19] is used iteratively to find the fixed points of Eq. (1.18), and then the Almeida-Pineda method [20, 21] is used to compute the gradient to minimize the loss function. The loss function, for instance, could be the square loss between predicted values of the networks and ground-truth values.

While GNN† is conceptually significant, it comes with several limitations. For Equation (1.18) to possess a unique solution, $f(\cdot)$ must function as a "contraction map" [18], significantly constraining its modeling capabilities. Due to the necessity for numerous iterations between gradient descent steps, GNN incurs substantial computational expenses. Some variants are introduced to overcome these limitations. The Graph Echo State Network (GESNet) [22] is developed as an extension of echo state networks, aiming to enhance the training efficiency of GNN†. The encoder, initialized randomly and requiring no training, employs a contractive state transition function to iteratively update the states of the graph nodes until reaching the fixed node states.

The stochastic steady-state method [23] introduces a scalable learning algorithm tailored for large graphs. In this method, the way the node state is updated in each iteration is stochastic and asynchronous. To ensure stability, the stochastic steady-state method defines its recurrent function as a weighted average of historical and new states. Although conceptually significant, SSE lacks theoretical proof that the repeated application of the recurrent function will lead to the gradual convergence of node states to fixed points. Gated GNN [24] enhances GNN† by integrating several modifications. Notably, gated GNN uses a gated recurrent unit (GRU) [25] as the transition function f, eliminating the need for a "contraction map" and keeping a fixed number of iteration steps. Meanwhile, modern optimization techniques can be used by gated GNN. The authors suggest employing multiple networks operating sequentially to generate a sequence output, with potential applications in fields like program verification. CommNet [26] employs GNN to facilitate multi-agent communication in AI systems, treating each agent as a node and iteratively updating agent states through communication before actions are taken. Interaction Network [27] employs GNN for physical reasoning, depicting objects as nodes, relationships as edges, and simulating the process with a pseudo-time system. VAIN [28] introduces attention mechanisms to enhance CommNet and Interaction Network.

1.2.2 Graph Convolutional Network

Graph Convolutional Networks (GCNs) are a class of deep learning models for processing graph-structured data using convolution operation defined on graph structure. Entities in graph-structured data are represented as nodes, and relationships between entities are represented as edges. The goal of graph convolutional networks is to learn useful representations based on convolution operation from graph-structured data, and use these representations for various tasks, such as node classification, graph classification, link prediction, and so on.

At present, GCNs mainly have two operation methods: spectral convolution and spatial convolution. Among them, spectral convolution method is based on graph signal processing [29]. Early spectral convolution neural networks use RNN architecture to presuppose a static internal graph structure and use Laplacian matrix as the frequency domain representation of the graph [30]. A learnable set of parameters is used as a frequency domain filter for the convolution operation. Because the Eigen-value Decomposition (EVD) of Laplacian matrix is repeatedly performed in the iterative training process, the spectral convolution method has high operational complexity. To solve this problem, ChebyNet is proposed in literature [31], which uses Chebyshev polynomial approximation to replace EVD. Similarly, literature [32] further reduces the computational difficulty by using a local first-order approximation of spectral convolution. GCN based on spectral convolution generally builds neural networks on the basis of static graph structure, but the graph data in practical problems is usually dynamic, and the nodes or edges may appear or disappear, and the attributes of nodes or edges may also change with time. Therefore, the research of novel graph convolutional neural networks which can adapt to the dynamic changes of graph data is a hot topic nowadays. Spatial convolution is a simple operation method based on neighborhood nodes. Literature [33] proposes to use message passing network to summarize such models, arguing that spatial convolution consists of two processes: message passing and state update. It is also pointed out that there are similar message functions in GG-NN [24], Interaction Network [34], DTN [35] and many spectral convolutional networks [30–32]. Literature [36] proposes a graph attention network to improve message transmission through attention mechanism, so that the model pays more attention to the learning of key nodes. Therefore, compared with spectral convolution method, spatial convolution method has the characteristics of simple operation and suitable for various graph structures, and how to realize the effective learning of edge nodes is the key problem to be solved urgently. GCN using spatial convolution generally does not use RNN as an internal coding network, but uses a multi-layer convolution architecture similar to modern CNN, and combines activation functions to realize nonlinear feature mapping, improving feature abstraction ability layer by layer. In literature [32], a multilayer linear model is established by using the first-order approximation of Chebyshev polynomials, so that spectral convolution can be computed layer by layer like spatial convolution, so as to establish a multilayer depth map convolution model, which is a useful attempt to synthesize the two convolution methods. To sum up, there are still few methods that can give full play to the advantages of the two convolution methods at the same time. Therefore, in-depth exploration of spectral convolution and spatial convolution fusion methods will be one of the most valuable research directions.

1.2.3 Graph Attention Network

Nowadays, many interesting tasks involve data that present in an irregular domain, and such data are usually represented in the form of graphs. Some researchers have made several attempts to extend the successful convolutional neural networks to deal with those arbitrarily structured graphs. GNNs were introduced by Gori et al. [15] and further refined in the work of Scarselli et al. [37], and it can directly deal with a more general class of graphs. GNNs often involve an iterative process known as message passing or propagation, which allows the nodes in a graph to exchange information and update their states iteratively until a stable or equilibrium state is reached. The attention mechanism has demonstrated considerable success in numerous computer vision tasks. Inspired by those attention based works, the graph attention networks are proposed for high performance. Contrary to GCNs, which treat all neighbors of a node uniformly, the attention mechanism stands out by assigning distinct attention scores to each neighbor. This allows for the identification and prioritization of more salient or critical neighbors.

Velickovic et al. [36] introduce the Graph Attention Network (GAT), a pioneering model that integrates the attention mechanism into the propagation steps. We will start by describing a single graph attentional layer to present the GAT. To achieve sufficient expressive power for the transformation of input features into higher-level features, it is imperative to incorporate at least one learnable linear transformation. Here, a shared linear transformation, parametrized by a weight matrix, $W \in \mathbb{R}^{F' \times F}$, is applied to every node. Then, a self-attention mechanism is defined for the nodes, quantifying the attention coefficients for pairs of nodes through a shared attentional mechanism denoted as a. The relationship strength between node i and j can be denoted as following:

$$e_{ij} = a(W\vec{\mathbf{h}}_i, W\vec{\mathbf{h}}_j), \tag{1.20}$$

where $\vec{\mathbf{h}}_i$ and $\vec{\mathbf{h}}_j$ are the features of the nodes i and j. In order to ensure comparability of coefficients across different nodes, the attention coefficients are typically normalized using the softmax function:

$$\alpha_{ij} = \text{softmax}_j(e_{ij}) = \frac{\exp(e_{ij})}{\sum_{k \in N_i} \exp(e_{ik})}. \tag{1.21}$$

In the GAT, the attention mechanism a is defined as a single-layer feedforward neural network, comprising a linear transformation with the weight vector $\overrightarrow{a} \in \mathbb{R}^{2F'}$ and a LeakyReLU nonlinear activation function with a negative input slope $\alpha = 0.2$. Then, the relationship computed by the attention mechanism can be expressed as:

$$\alpha_{ij} = \frac{\exp(\text{LeakyReLU}(\overrightarrow{a}^T[W\overrightarrow{\mathbf{h}}_i||W\overrightarrow{\mathbf{h}}_j]))}{\sum_{k \in N_i} \exp(\text{LeakyReLU}(\overrightarrow{a}^T[W\overrightarrow{\mathbf{h}}_i||W\overrightarrow{\mathbf{h}}_k]))}, \tag{1.22}$$

where $\overrightarrow{a}^T$ represents transposition and $||$ denotes the concatenation operation.

Applying the self-attention strategy, GAT computes the hidden state of each node by attentively considering its neighbors. Hence, it is evident that the GAT exhibits superior performance compared to GCN across multiple tasks, encompassing semi-supervised node classification and link prediction. Conventional aggregation functions such as mean or sum, although standard and well-examined, possess limited capabilities. They lack the ability to effectively capture interactions among neighboring nodes. Michail et al. [38] formalize these interactions using an information-theoretic framework that notably includes synergistic information, and introduce the Graph Ordering Attention (GOAT) layer, a novel GNN component that captures interactions between nodes in a neighborhood. However, both the GCNs and GAT have shown the strengths and weaknesses of the resulting GNN architectures in recent works. Hence, aiming at exploiting the strengths of both approaches to their full extent, Javaloy et al. [39] introduce the graph convolutional attention layer (CAT), which relies on convolutions to compute the attention scores. Based on the CAT, they design the learnable graph convolutional attention network (L-CAT), which is composed of a GNN architecture that automatically interpolates between GCN, GAT and CAT in each layer. For heterogeneous graph, the methods usually fail to represent the non-pairwise relations in heterogeneous graph, or only capable of capturing local information around target node. To handle these problem, Liu et al. [40] propose a metapath based multilevel graph attention networks (MMAN) to jointly learn node embeddings on two substructures, i.e., metapath based graphs and hypergraphs extracted from original heterogeneous graph. In addition, the GAT is widely used in some downstream tasks. According to specific tasks, GAT is optimized to obtain a higher performance model. Zhang et al. [41] propose a Metadata Neighbourhood Graph Co-Attention Network (MangoNet) to model the correlations between each target image and its neighbours. To accurately capture the visual clues from the neighbourhood, a co-attention mechanism is introduced to embed the target image and its neighbours as graph nodes, while the graph edges capture the node pair correlations. Wang et al. [42] propose a Type-Aware Graph Attention Networks (TAGAT) for the reasoning task over knowledge graphs (KGs). Except for combining type-related information during the embedding process, TAGAT further adopts a hierarchical attention mechanism to realize the perception of type and neighborhood information meticulously. For person reidentification (Re-ID), the traditional methods exclusively focus on establishing intra-local relations among parts within a single pedestrian image. However, they overlook the crucial inter-local relations among parts originating from different images, leading to an incomplete representation of local relation information. Zhang et al. [43] propose a novel deep graph model named Heterogeneous Local Graph Attention Networks (HLGAT) to model the inter-local relation and the intra-local relation in the completed local graph, simultaneously. Concretely, they establish a comprehensive local graph using local features. Subsequently, they employ the attention mechanism for aggregating these local features, thereby emphasizing their individual significance during the learning process of inter-local and intra-local relations.

1.2.4 Frontier of Graph Neural Networks

Generative Models for Graphs. Graph generation is a pivotal challenge within the realm of graph representation learning, involving the task of understanding the distributions of provided graphs and generating new graph structures similar to real-world graph data. Graph generation can be applied in fields such as social networks, protein structure modeling, drug discovery, anomaly detection, and scene understanding. The Erdos-Renyi model [44], stochastic block models [1], the small-world model [45], and the preferential attachment model [46] are representative traditional methods for graph generation. The Erdos-Renyi model stands out as one of the most basic and widely recognized generative graph models. The Erdos-Renyi model posits that the likelihood of an edge forming between any node pairs is uniform and represented by a specific parameter. As these edge probabilities are independent of each other, the computational time complexity for generating a graph scales linearly with the size of the adjacency matrix. The pivotal advancement in stochastic block models lies in their capacity to manipulate edge probabilities both within and between distinct blocks. The preferential attachment model produces graphs featuring power-law degree distributions through an iterative approach.

The conventional methods for generating graphs prove effective in creating synthetic graphs with specific characteristics and offer valuable insights into the potential emergence of particular graph structures in real-world scenarios. Nevertheless, a significant drawback of these traditional approaches lies in their dependence on a predetermined, hand-crafted generation process. They lack the capability to autonomously learn to generate graphs with some characteristics from data. The Generative Adversarial Network (GAN), Variational Autoencoder (VAE), and Autoregressive model (AR) are three widely used approaches to build deep generative models. Recently, Normalizing Flow (NF) and Diffusion Model (DM) are also used for graph generation. Reference [47] adopts the GAN framework for the generation of molecular graphs. A molecular graph, denoted as G with N nodes, is represented using $\mathbf{F}$ and $\mathbf{E}$, where $\mathbf{F}$ and $\mathbf{E}$ indicate the type of all nodes and the type of all edges respectively.

An additional component, referred to as the judge, is introduced. The judge assesses the quality of a generated graph based on the specified property, providing a measure of how well the generated graph aligns with the desired characteristics. Simonovsky et al. propose a graph-level VAE termed GraphVAE [48]. The authors employ two separate Graph Neural Networks (GNNs) to parameterize both the mean and variance of a posterior normal distribution concerning latent variables. GraphVAE integrates a fundamental Multi-Layer Perceptron (MLP) with a Bernoulli distributional assumption. GraphVAE specifies a particular node ordering function as part of the approach. Put differently, the overall objective of maximizing log-likelihood is then akin to employing a set of independent binary cross-entropy loss functions, one for each edge. Readers can refer to [49] to learn more about deep graph generation models.

Dynamic Graph Neural Networks. Dynamic Graph Neural Networks (DGNNs), which nodes and edges can be added or removed dynamically over time, are used in a wide range of fields, including social networks, biological networks, transportation systems and scene understanding tasks. Hence, DGNNs will have a large and far-reaching impact on various fields. DGNNs are designed to encode both structural and temporal patterns within dynamic networks. For capturing structural patterns, DGNNs commonly leverage GNNs. On the other hand, to address temporal patterns, DGNNs typically incorporate time series modules like Recurrent Neural Networks (RNNs).

Traditional Graph Convolutional Networks (GCNs) face limitations imposed by fixed graph structures, hindering their ability to adaptively integrate node features for comprehensive information representation. To overcome this constraint, Li et al. [50] introduced a Generalized GCN that operates on arbitrary graph structures. This innovation allows GCNs to dynamically construct task-driven adaptive graphs for individual training samples, providing a more flexible and tailored approach to graph-based learning. The modeling of intricate spatial and temporal correlations within correlated time series data is crucial for comprehending the dynamics of traffic and forecasting the future state of an evolving traffic system. Toward this, Bai et al. [51] propose two adaptive modules for enhancing Graph Convolutional Network (GCN) with new capabilities: (1) a Node Adaptive Parameter Learning (NAPL) module to capture node-specific patterns; (2) a Data Adaptive Graph Generation (DAGG) module to infer the inter-dependencies among different traffic series automatically. Zhang et al. [52] introduce a novel method called Dynamic Graph Neural Network for Sequential Recommendation (DGSR). This approach connects different user sequences through a dynamic graph structure, enabling the exploration of interactive behavior among users and items while considering time and order information. Further, Zhu et al. [53] present the WinGNN framework for modeling dynamic graphs. This is achieved through a simple Graph Neural Network (GNN) model employing a meta-learning strategy and a novel mechanism called random gradient aggregation. WinGNN computes the frame-wise loss of the current snapshot, and the loss gradient is then passed to the next snapshot to model graph dynamics without the need for temporal encoders.

Neural Architecture Search has become a research hotspot in deep learning to solve the architecture design problem of graph neural networks. The neural network architecture search method has three key elements: search space, search strategy, and performance evaluation strategy. In early research, neural network architecture search methods using reinforcement learning search strategies [54] generally used discrete search spaces and typically required training an RNN controller to generate candidate architectures. The advantage of this type of method is that it conforms to the discrete structure of neural network architecture, but the disadvantage is that it requires thousands of GPU days of computation [54]. The search strategy based on evolutionary algorithms [55] can reduce the computational load to several thousand GPUs by generating a small number of candidate models. In addition, during the search process, exploring new architectures faces new performance uncertainties. If existing architectures are fully utilized, it is necessary to determine which nodes are under trained and continue training. The search strategy is required to have a certain trade-off mechanism, which is the "exploration exploitation dilemma" faced by most neural network architecture search methods, including those based on reinforcement learning and evolutionary algorithms.

In recent years, the establishment of continuous search spaces [56] has provided new ideas for search strategies. Literature [56] proposes a differentiable architecture search method, which uses gradient methods to jointly optimize the implicit weights and architecture distribution parameters within the network model in a continuous search space. Although relaxing to a continuous search space can accelerate the search through gradient descent, the resulting architecture still needs to be discretized, which may result in the final architecture not belonging to the original discrete search space. Therefore, how to introduce discrete constraints while ensuring the solvability of joint optimization problems is a key issue that urgently needs to be studied in the current search methods for continuous space neural network architectures. For automated search of graph neural network model architecture, using architecture search technology based on graph neural network units is an effective means. For example, to solve the problem of action recognition based on human skeleton data, literature [57] proposes a Cross Entropy method with Import Mixing search strategy for graph convolutional neural network models, which establishes a discrete search space with 8 predefined basic units. By searching for the optimal combination of graph neural network modules at different levels of abstract representation, a graph neural network architecture is established.

1.3 Chapter Outline

This book delves deeply into the application and development of Graph Neural Networks (GNNs) for tasks related to scene understanding. It aims to introduce the reader to the pivotal concepts, challenges, and application of graph neural networks in scene understanding. Serving as a gateway to the subsequent chapters, this chapter provides a panoramic view of the book. The subsequent chapters delve into specific aspects, methodologies, and applications of GNNs in scene understanding, and the details of each chapter are as follows.

In this chapter, we provide an extensive introduction to the world of graph representation learning and graph neural networks. We begin with a deep dive into graph representation learning, where we unravel the essential knowledge needed to understand this field. Starting with the fundamental concepts of graphs, we proceed to explore their statistical properties. We then delve into the spectral methods applied to graphs, laying the groundwork for understanding frequency-domain graph convolutional networks. Continuing our exploration, we introduce early graph representation learning techniques, particularly focusing on two main categories: matrix decomposition and random walk based methods. Transitioning to the world of graph neural networks, we initially introduce the original graph neural network, which employs uniform parameters across all layers and updates node states through iterative aggregation. We then shift our attention to popular graph convolutional networks. Our exploration culminates with graph attention networks, which offer the flexibility to adjust edge weights. Finally, we offer a glimpse into recent advancements in graph neural networks.

In Chap. 2, we mainly introduce scene understanding, which is one of the most challenging and important problems in computer vision in recent years. In this chapter, we first introduce the tasks of scene understanding. According to scene data type, the scene understanding can be categorized into image-based understanding, video-based understanding and 3D based understanding. Further, the tasks of scene understanding can be divided into several levels according to task target, as shown in Fig. 2.1. The lower-level (level-1 and level-2, L1 & L2) tasks which process single-label or multi-label category are easy, and higher-level (level-3, L3) tasks which process detection or localization become a little harder. At last, the highest-level (level-4, L4) tasks which are also the hardest, are used to process semantic analysis and generation.

In Chap. 3, we mainly introduce basic theory of graph neural network. We first introduce the general framework of graph neural networks, which mainly consists of feature transformation layers, activation layers, and pooling layers. The feature transformation layer can be further divided into aggregation step and combination step. Then, we introduce the typical operations and representative methods of graph neural networks respectively. For graph neural network operations, we describe spectral graph convolution, spatial graph convolution, graph attention, and graph pooling. For graph neural network architectures, we present representative methods for vanilla graph neural networks, graph convolutional networks, and graph attention networks, respectively, enabling readers to grasp the details of methods for different types of graph neural networks. Finally, we summarize the training methods for graph neural networks.

In Chap. 4, we mainly introduce how to solve the problem of multi-scale objects in scene analysis, a self-supervised feature fusion convolutional network model is proposed to improve the resolution accuracy of multi-scale objects in field images. Based on the graph convolutional network, SFGCN designs and adds K-level neighborhood graph spatial convolution, spectral graph convolution feature fusion and self-supervised attention, scene multi-domain adaptive pooling and other modules to strengthen the model's feature learning ability for local small-scale targets and realize pixel level semantic label parsing of scene images. Firstly, the model uses hierarchical grid division to construct the feature data of scene graph, and then constructs the network model based on the spatial graph convolution operation of K-level neighborhood to realize the global feature learning. Then, to solve the problem that local semantic labels are difficult to parse, a graph attention module based on self-supervised feature fusion is proposed. By combining spectral convolution and attention mechanism, self-supervised information is added to local model weight training to enhance local feature learning. Finally, in order to maintain the intra-domain consistency of multi-domain scene images, a multi-domain adaptive scene pooling method is proposed to improve the intra-class inconsistency caused by the differences of scene images.

In Chap. 5, we mainly introduce a graph neural network for supervised human parsing and a graph neural network for weakly supervised human parsing. Human body data is inherently complex and can be organized in hierarchical structures. In this chapter, we construct a spatial graph and a semantic graph for human parsing, which can provide a structured way to model and analyze the human body. Further, we argue that the hierarchical structure of human body helps to improve the accuracy of human parsing model. Hence, this chapter proposes a method which takes full advantage of the hierarchical properties of the human body and the strengths of GNNs. Considering the cost of pixel-level annotations, we try to explore the possibility of graph neural network model in weakly supervised human parsing. With the box-level labels and point-level labels, we propose a two-stage learning framework, by combining generation pseudo labels and graph neural network based weakly supervised human parsing.

In Chap. 6, we mainly introduce the dynamic graph neural networks for human parsing. The adaptive graph structure allows the framework to effectively handle the evolving entities and relations, making it suitable for scenarios where the underlying data is subject to changes and updates over time. For human parsing, it is common that a training image sample contains multiple classes, but not all of the labeled classes in the dataset are present in the image. To overcome the limitations of the fixed graph, we are making efforts to design a more accurate and optimal graph for this specific training sample, and introduce a novel framework called multilabel learning based adaptive graph convolutional network (ML-AGCN). This framework can dynamically generate adaptive graphs for various training samples by leveraging the principles of multilabel learning. Further, we propose a dynamic graph pooling for human parsing. By generating the dynamic hierarchical semantic graph in a dense-to-coarse manner, we naturally capture and utilize the structural information. It enables the model to dynamically adjust the receptive fields during the pooling process, which is particularly crucial for capturing both local and global context in an image.

In Chap. 7, we introduce graph neural networks to video object segmentation. To capture the structural shape information of the target, we first propose a graph structure generation algorithm based on agglomerative clustering. As to graph construction for video data, we use the minimum spanning tree as the generated graph structure, which allows nodes with similar features to be connected, thus effectively capturing the structural shape information of the target. To further enhance the target structural features, we introduce a graph neural network to accomplish iterative message passing, while combining the structural feature extraction module with the visual Transformer can obtain more expressive target features. In order to reduce the prediction noise caused by the interference of similar objects, we design a prototype matching mechanism. First, we construct a prototype feature extraction algorithm based on Gaussian mixture model to represent the whole target with a small number of prototype features. Then, we perform prototype feature matching to enhance the robustness of the feature matching process. In addition, we add object-level matching loss to the total loss function to introduce more supervised information for network training.

In Chap. 8, we mainly make summary and prospect, which summarizes the work of this book, and puts forward some ideas for the future research direction. The book focuses on graph neural network methods and applications for scene understanding. Graph neural network is an important method for graph data processing. Scene understanding is one of the focuses in computer vision research, and key technology to realize semantic segmentation and recognition of image or video. In future work, we mainly discuss some scene understanding tasks that could be solved or improved by graph neural network. Some of those tasks have been studied with graph representation and graph model, and have achieved some progress, then others could be potential research direction. The goal of this book is to bring, in a timely fashion, the latest advances and developments in graph neural network, and scene understanding methods based on graph neural network, which is particularly suited for readers who are interested in the research and technology innovation in machine learning and computer vision. Features of the book include graph convolutional network based on self-supervised feature fusion is designed for scene parsing, adaptive graph convolutional network based on multi-label learning is designed for human parsing, and graph construction and graph neural network architecture for video object segmentation are proposed.

References

1. Newman, M.: Networks. Oxford University Press (2018)
2. Weisfeiler, B., Lehman, A.: A reduction of a graph to a canonical form and an algebra arising during this reduction. Nauchno-Technicheskaya Informatsia 2(9), 12–16 (1968)
3. Shervashidze, N., Schweitzer, P., van Leeuwen, E.J., Mehlhorn, K., Borgwardt, K.M.: Weisfeiler-lehman graph kernels. The Journal of Machine Learning Research 12, 2539–2561 (2011)
4. Kashima, H., Tsuda, K., Inokuchi, A.: Marginalized kernels between labeled graphs. In: Proceedings of the International Conference on Machine Learning, pp. 321–328 (2003)
5. Borgwardt, K.M., Kriegel, H.: Shortest-path kernels on graphs. In: Proceedings of the IEEE International Conference on Data Mining, pp. 74–81 (2005)

6. Lü, L., Zhou, T.: Link prediction in complex networks: A survey. Physica A: Statistical Mechanics and its Applications 390(6), 1150–1170 (2011)
7. Hamilton, W.L.: Graph representation learning. Morgan & Claypool Publishers (2020)
8. Ahmed, A., Shervashidze, N., Narayanamurthy, S.M., Josifovski, V., Smola, A.J.: Distributed large-scale natural graph factorization. In: Proceedings of the International Conference on World Wide Web, pp. 37–48 (2013)
9. Ou, M., Cui, P., Pei, J., Zhang, Z., Zhu, W.: Asymmetric transitivity preserving graph embedding. In: Proceedings of the International Conference on Knowledge Discovery and Data Mining, pp. 1105–1114 (2016)
10. Cao, S., Lu, W., Xu, Q.: Grarep: Learning graph representations with global structural information. In: Proceedings of the International Conference on Information and Knowledge Management, pp. 891–900 (2015)
11. Perozzi, B., Al-Rfou, R., Skiena, S.: Deepwalk: online learning of social representations. In: Proceedings of the International Conference on Knowledge Discovery and Data Mining, pp. 701–710 (2014)
12. Grover, A., Leskovec, J.: Node2vec: Scalable feature learning for networks. In: Proceedings of the International Conference on Knowledge Discovery and Data Mining, pp. 855–864 (2016)
13. Tang, J., Qu, M., Wang, M., Zhang, M., Yan, J., Mei, Q.: LINE: large-scale information network embedding. In: Proceedings of the International Conference on World Wide Web, pp. 1067–1077 (2015)
14. Qiu, J., Dong, Y., Ma, H., Li, J., Wang, K., Tang, J.: Network embedding as matrix factorization: Unifying deepwalk, line, pte, and node2vec. In: Proceedings of the International Conference on Web Search and Data Mining, pp. 459–467 (2018)
15. Gori, M., Monfardini, G., Scarselli, F.: A new model for learning in graph domains. In: Proceedings of IEEE International Joint Conference on Neural Networks, vol. 2, pp. 729–734 (2005)
16. Scarselli, F., Tsoi, A.C., Gori, M., Hagenbuchner, M.: Graphical-based learning environments for pattern recognition. In: Structural, Syntactic, and Statistical Pattern Recognition, Joint IAPR International Workshops, vol. 3138, pp. 42–56 (2004)
17. Scarselli, F., Gori, M., Tsoi, A.C., Hagenbuchner, M., Monfardini, G.: The graph neural network model. IEEE Transactions on Neural Networks 20(1), 61–80 (2009)
18. Khamsi, M.A., Kirk, W.A.: An introduction to metric spaces and fixed point theory. John Wiley & Sons (2011)
19. Powell, M.J.D.: An efficient method for finding the minimum of a function of several variables without calculating derivatives. The Computer Journal 7(2), 155–162 (1964)
20. Almeida, L.B.: A learning rule for asynchronous perceptrons with feedback in a combinatorial environment. In: IEEE International Joint Conference on Neural Networks, vol. 2, pp. 609–618 (1987)
21. Pineda, F.J.: Generalization of back-propagation to recurrent neural networks. Physical Review Letters 59(19), 2229–2232 (1987)
22. Gallicchio, C., Micheli, A.: Graph echo state networks. In: International Joint Conference on Neural Networks, pp. 1–8 (2010)
23. Dai, H., Kozareva, Z., Dai, B., Smola, A.J., Song, L.: Learning steady-states of iterative algorithms over graphs. In: Proceedings of the International Conference on Machine Learning, vol. 80, pp. 1114–1122 (2018)
24. Li, Y., Tarlow, D., Brockschmidt, M., Zemel, R.S.: Gated graph sequence neural networks. In: Proceedings of the International Conference on Learning Representations (2016)
25. Cho, K., van Merrienboer, B., Gülçehre, Ç., Bahdanau, D., Bougares, F., Schwenk, H., Bengio, Y.: Learning phrase representations using RNN encoder-decoder for statistical machine translation. In: Proceedings of the Conference on Empirical Methods in Natural Language Processing, pp. 1724–1734 (2014)
26. Sukhbaatar, S., Szlam, A., Fergus, R.: Learning multiagent communication with backpropagation. In: Advances in Neural Information Processing Systems, pp. 2244–2252 (2016)

27. Battaglia, P.W., Pascanu, R., Lai, M., Rezende, D.J., Kavukcuoglu, K.: Interaction networks for learning about objects, relations and physics. In: Advances in Neural Information Processing Systems, pp. 4502–4510 (2016)
28. Hoshen, Y.: VAIN: attentional multi-agent predictive modeling. In: Advances in Neural Information Processing Systems, pp. 2701–2711 (2017)
29. Shuman, D.I., Narang, S.K., Frossard, P., Ortega, A., Vandergheynst, P.: The emerging field of signal processing on graphs: Extending high-dimensional data analysis to networks and other irregular domains. IEEE Signal Processing Magazine 30(3), 83–98 (2013)
30. Bruna, J., Zaremba, W., Szlam, A., LeCun, Y.: Spectral networks and locally connected networks on graphs. In: International Conference on Learning Representations (2014)
31. Defferrard, M., Bresson, X., Vandergheynst, P.: Convolutional neural networks on graphs with fast localized spectral filtering. In: Advances in Neural Information Processing Systems, vol. 29, pp. 3844–3852 (2016)
32. Kipf, T.N., Welling, M.: Semi-supervised classification with graph convolutional networks. In: International Conference on Learning Representations (2017)
33. Gilmer, J., Schoenholz, S.S., Riley, P.F., Vinyals, O., Dahl, G.E.: Neural message passing for quantum chemistry. In: International Conference on Machine Learning, vol. 70, pp. 1263–1272 (2017)
34. Battaglia, P.W., Pascanu, R., Lai, M., Rezende, D.J., Kavukcuoglu, K.: Interaction networks for learning about objects, relations and physics. In: Advances in Neural Information Processing Systems, pp. 4502–4510 (2016)
35. Schütt, K.T., Arbabzadah, F., Chmiela, S., Müller, K.R., Tkatchenko, A.: Quantum-chemical insights from deep tensor neural networks. Nature Communications 8(1) (2017)
36. Velickovic, P., Cucurull, G., Casanova, A., Romero, A., Liò, P., Bengio, Y.: Graph attention networks. In: International Conference on Learning Representations (2017)
37. Scarselli, F., Gori, M., Tsoi, A.C., Hagenbuchner, M., Monfardini, G.: The graph neural network model. IEEE Transactions on Neural Networks 20(1), 61–80 (2009)
38. Chatzianastasis, M., Lutzeyer, J.F., Dasoulas, G., Vazirgiannis, M.: Graph ordering attention networks. In: Proceedings of the AAAI Conference on Artificial Intelligence, pp. 7006–7014 (2023)
39. Javaloy, A., Sánchez-Martín, P., Levi, A., Valera, I.: Learnable graph convolutional attention networks. In: International Conference on Learning Representations (2023)
40. Liu, J., Song, L., Gao, L., Shang, X.: Mman: metapath based multi-level graph attention networks for heterogeneous network embedding (student abstract). In: Proceedings of the AAAI Conference on Artificial Intelligence, pp. 13005–13006 (2022)
41. Zhang, J., Wu, Q., Zhang, J., Shen, C., Lu, J.: Mind your neighbours: image annotation with metadata neighbourhood graph co-attention networks. In: Proceedings of the IEEE Conference on Computer Vision and Pattern Recognition, pp. 2956–2964 (2019)
42. Wang, Y., Wang, H., He, J., Lu, W., Gao, S.: Tagat: type-aware graph attention networks for reasoning over knowledge graphs. Knowl. Based Syst. 233, 107500 (2021)
43. Zhang, Z., Zhang, H., Liu, S.: Person re-identification using heterogeneous local graph attention networks. In: Proceedings of the IEEE Conference on Computer Vision and Pattern Recognition, pp. 12136–12145 (2021)
44. Erdös, P., Rényi, A.: On the evolution of random graphs. Publication of the Mathematical Institute of the Hungarian Academy of Sciences 5(1), 17–61 (1960)
45. Watts, D.J., Strogatz, S.H.: Collective dynamics of 'smallworld'networks. Nature 393(6684), 440–442 (1998)
46. Albert, R., Barabási, A.L.: Statistical mechanics of complex networks. Reviews of Modern Physics 74(1), 47–97 (2002)
47. De Cao, N., Kipf, T.: MolGAN: An implicit generative model for small molecular graphs. arXiv:1805.11973 (2018)
48. Simonovsky, M., Komodakis, N.: Graphvae: Towards generation of small graphs using variational autoencoders. In: Proceedings of the International Conference on Artificial Neural Networks, pp. 412–422 (2018)

49. Guo, X., Zhao, L.: A systematic survey on deep generative models for graph generation. IEEE Transactions on Pattern Analysis and Machine Intelligence 45(5), 5370–5390 (2023)
50. Li, R., Wang, S., Zhu, F., Huang, J.: Adaptive graph convolutional neural networks. In: Proceedings of the Thirty-Second AAAI Conference on Artificial Intelligence, the 30th innovative Applications of Artificial Intelligence, and the 8th AAAI Symposium on Educational Advances in Artificial Intelligence, pp. 3546–3553 (2018)
51. Bai, L., Yao, L., Li, C., Wang, X., Wang, C.: Adaptive graph convolutional recurrent network for traffic forecasting. In: H. Larochelle, M. Ranzato, R. Hadsell, M. Balcan, H. Lin (eds.) Advances in Neural Information Processing Systems (2020)
52. Zhang, M., Wu, S., Yu, X., Liu, Q., Wang, L.: Dynamic graph neural networks for sequential recommendation. IEEE Transactions On Knowledge And Data Engineering 35(5), 4741–4753 (2023)
53. Zhu, Y., Cong, F., Zhang, D., Gong, W., Lin, Q., Feng, W., Dong, Y., Tang, J.: Wingnn: Dynamic graph neural networks with random gradient aggregation window. In: Proceedings of the 29th ACM Conference on Knowledge Discovery and Data Mining, pp. 3650–3662. ACM (2023)
54. Zoph, B., Le, Q.V.: Neural architecture search with reinforcement learning. In: International Conference on Learning Representations (2017)
55. Real, E., Moore, S., Selle, A., Saxena, S., Leon-Suematsu, Y.I., Tan, J., Le, Q.V., Kurakin, A.: Large-scale evolution of image classifiers. In: International Conference on Machine Learning, vol. 70, pp. 2902–2911 (2017)
56. Liu, H., Simonyan, K., Yang, Y.: DARTS: differentiable architecture search. In: International Conference on Learning Representations (2019)
57. Peng, W., Hong, X., Chen, H., Zhao, G.: Learning graph convolutional network for skeleton-based human action recognition by neural searching. In: AAAI Conference on Artificial Intelligence, pp. 2669–2676 (2020)

Chapter 2
Scene Understanding

A scene could be a natural area or a place of human-made where human being could act within. Usually, the man-made scene is divided into indoor scene and outdoor scene. Since places, objects and events that make up a scene are usually complex and indistinct, it is difficult to directly learn intensional pattern of the whole scene. Scene understanding is a kind of computer vision task to get category of place, structure of object, attribute of event and other semantic information from the scene.

2.1 Overview of Scene Understanding

The primary scene understanding tasks include scene classification, scene attribute recognition and so on. The scene classification is to predict category name of a given scene image by pattern classification model, like SVM, Bayes, CNN, then feature extraction and fusion algorithms are studied to convert the original image data into feature vector. This task can provide basic object and location context information of man-made or natural scene for many intelligent applications such as smart city, autonomous vehicles, and intelligent personal assistant. Not like scene classification task which outputs only one label of scene category, scene attribute is a multi-label classification task that recognizes multiple attributes of objects and background environment in specific scene. For example, in a mountain scene with clouds and natural light, there could be green grass, blue sky. Green, blue, natural are adjective attributes of the scene, and clouds, light, grass, sky are noun attributes. In most modern intelligent systems of intelligent transportation [1], intelligent vehicles [2], autonomous robot [3] etc., scene attribute recognition can help them make appropriate and reasonable decisions. And it is also an important basic subject for further research in the tasks of scene image caption [4], scene semantic segmentation [5] and scene understanding [6].

W. Liu et al., *Graph Neural Network Methods and Applications in Scene Understanding*,
https://doi.org/10.1007/978-981-97-9933-6_2

2.1.1 Data Types of Scene Understanding

Due to the diversity of data acquisition equipment, scene understanding has a variety of input data sources. Different scene data types are oriented to different scene understanding tasks. Concretely, according to scene data type, it can categorize scene understanding into image-based understanding, video-based understanding and 3D based understanding.

Usually, image-based scene understanding is a fundamental task which deals with RGB image captured by general digital visual sensor, such as pinhole camera, fish-eye camera and so on. These tasks aim to obtain information about objects and understand the scene only with the input RGB images. In addition, since RGB images are the easiest to capture and do not require much computing resources to process, image-based scene understanding is the most common computer vision task. With the continuous development of data acquisition equipment, the same scene can be expressed by multiple data sources. Hence, some tasks process hyper-spectral, infrared, thermal images by multi-sensor fusion method.

Different from image-based scene understanding, video-based scene understanding task is used to analyze continuous image frames, and each frame has a unique timestamp. Thus, these data are often constrained by strict time sequential order. How to correctly process and utilize the temporal relationship between video frames is very important for the task of video-based scene understanding. Video data is usually from digital video recorder and there are also formats of RGB, hyperspectral, infrared thermal, etc. With the rapid development of autonomous driving in recent years, video-based scene understanding has attracted the attention of many researchers and has gradually become a key problem of computer vision.

Finally, for 3D-based scene understanding, the 3D data is from not only visual sensor such as stereo camera but also laser LiDAR, inertial measurement unit and so on. Their formats include RGB-D data, LiDAR data, point-cloud data. There are different tasks for those three kinds of scene understanding. Actually, there are also some multi-data fusion-based scene understanding methods that take advantage of two or three data types to achieve robust performance. In general, we think that image-based scene understanding task is a little easier than video-based scene understanding task, while the 3D-based scene understanding task is much harder.

2.1.2 Task Levels of Scene Understanding

Besides, from easy to hard, the tasks of scene understanding can also be divided into several levels, as shown in Fig. 2.1. The lower-level (level-1, L1) tasks which process scene understanding tasks without semantic information are easy. Then higher-level (level-2, L2) tasks orient weak semantic scene understanding, which process detection or localization, become a little harder. At last, the highest-level (level-3, L3) tasks which are also the hardest, are used to process strong semantic scene understanding.

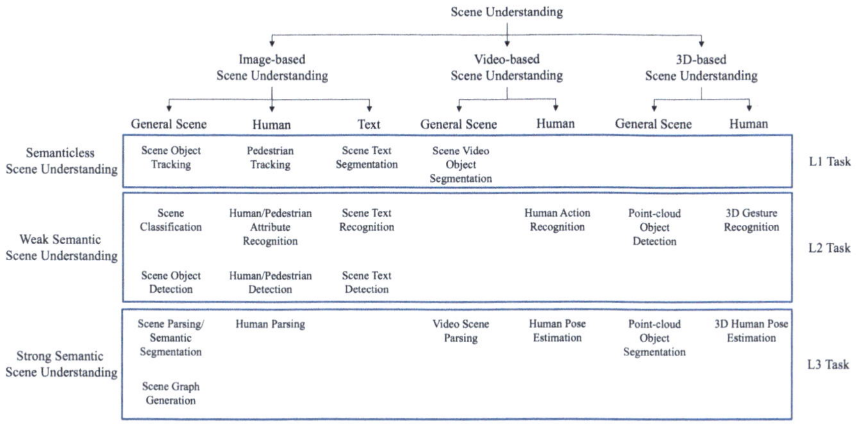

Fig. 2.1 Data types and task levels of scene understanding

As shown in Fig. 2.1, tasks become harder and harder from left to right and from top to bottom. For example, in image-based scene understanding, the scene object tracking is an L1 task since it does not need to learn the semantic information. The scene video classification is an L2 task, since it usually predicts single or multiple labels of one given video. While video object segmentation is an L3 task, because it must achieve semantic level feature learning to extract accurate object area from video. Besides, the outputs of L1 tasks are coarse grained. For example, the numerical representation of a scene category name for scene classification task is usually a scalar with only one dimension. And the scene understanding model of L1 predicts category name or attribute names for the whole scene image (or video) data. Thus, the output of L1 task is image-level (or video-level). Then for L2 tasks, their outputs become finer grained. For example, scene attribute localization can predict a bounding box of each scene attribute. It means that this scene understanding model must learn not only what the scene attribute names are, but also where the attributes show in the scene. Showing in Fig. 2.2, there are four scene images and every scene has two or three attributes. For example, Fig. 2.2a is a field scene, there are three attributes of blue-sky, rocky-mountain, green-filed. Figure 2.2b is a highway scene, there are also three attributes of overcast-sky, rocky-mountain, road. Note that, different scenes could have same attributes. Besides, scene object detection is similar with the scene attribute localization. It focuses on physical entities such as man-made things, human, animals and plants. The detection model predicts a box that contains the object and outputs the most possible object name. Four scene images from thermal camera are shown in Fig. 2.3, the scene object detection model predicts objects of person, smoke, train and car in those scenes and selects their specific regions.

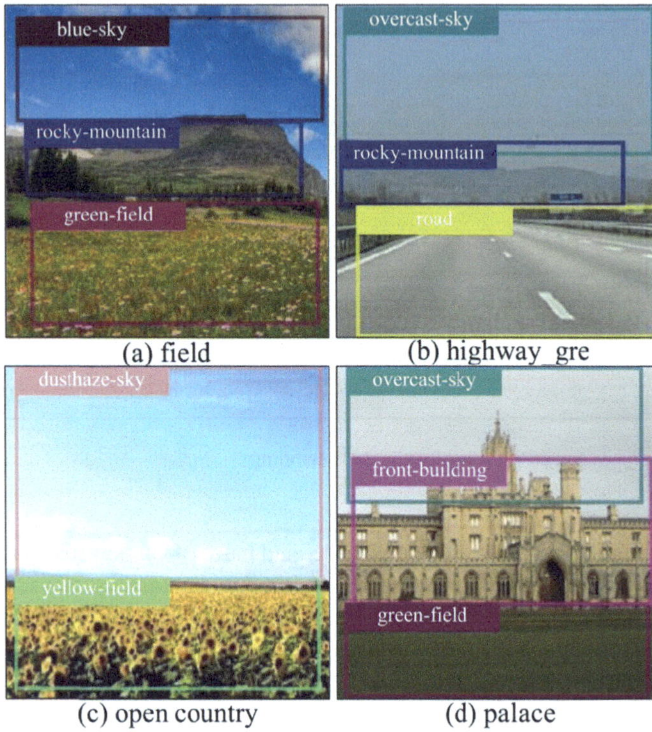

Fig. 2.2 Scene attribute localization

2.2 Taxonomy of Scene Understanding

For general scene understanding, according to understanding level, we categorize tasks of scene understanding into four classes, including scene representation, scene analysis, weak semantic scene understanding and strong semantic scene understanding, as shown in Fig. 2.4. The scene representation focuses on feature extraction and selection of scene data, which is a low-level processing. The scene analysis is mainly to do segmentation or object tracking on image or video scene data, which is usually without semantic information. For scene understanding tasks, we divide them into two levels according to complexity of semantic information. The weak semantic-oriented scene understanding includes tasks which only output coarse-grain and low-level semantic label, such classification label, detection bounding box. For example, the target of human action recognition task is to predict an action label for the given human action video, and the pedestrian detection is to localize the walking human in the image or video with a bounding box. The strong semantic-oriented scene understanding includes tasks which output fine-grain and high-level semantic label, such as classification label of each pixel, nodes label with logic connection.

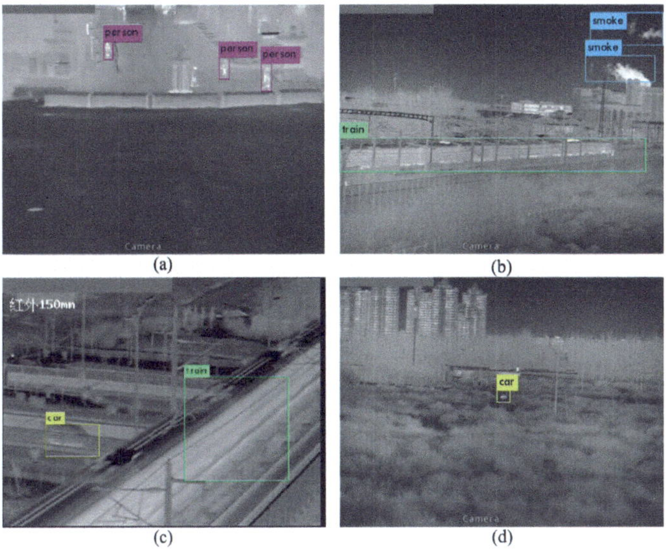

Fig. 2.3 Scene object detection

For example, scene parsing can predict pixel-level semantic label, and human pose estimation outputs skeleton with structure information of human body.

According to task target, each kind of scene understanding can be divided into many small tasks, we mainly consider general scene, human, text. For example, when taking the general scene as target, the image-based scene understanding usually processes scene classification, scene attribute recognition, scene attribute localization, scene object detection, scene parsing (also named semantic segmentation/understanding), scene graph generation, and interactive object segmentation. Then, for human target, the image-based scene understanding need process human parsing, human/pedestrian attribute recognition, human/pedestrian detection. In some city or road scene, text detection, text recognition and text segmentation are also typical tasks of image-based scene understanding. For video-based scene understanding, the tasks of general scene include scene video classification and video object segmentation (or video scene parsing). Besides, there are more tasks for human, such as human action recognition, human pose estimation, person re-identification and so on. Finally, based on 3D data, there are more tasks on general scene including point-cloud object classification, point-cloud object detection, point-cloud object segmentation, point-cloud generation, 3D scene reconstruction, visual event recognition. And there are also a few tasks of human target, such as 3D human pose estimation and 3D gesture recognition. Generally, the scene understanding task focuses on macroscopic objects, places, events in natural or man-made scene. Microscopic or local things are usually not under consideration. For example, the bio-metric recognition tasks such as face, finger print, palm, emotion, facial expression recognition

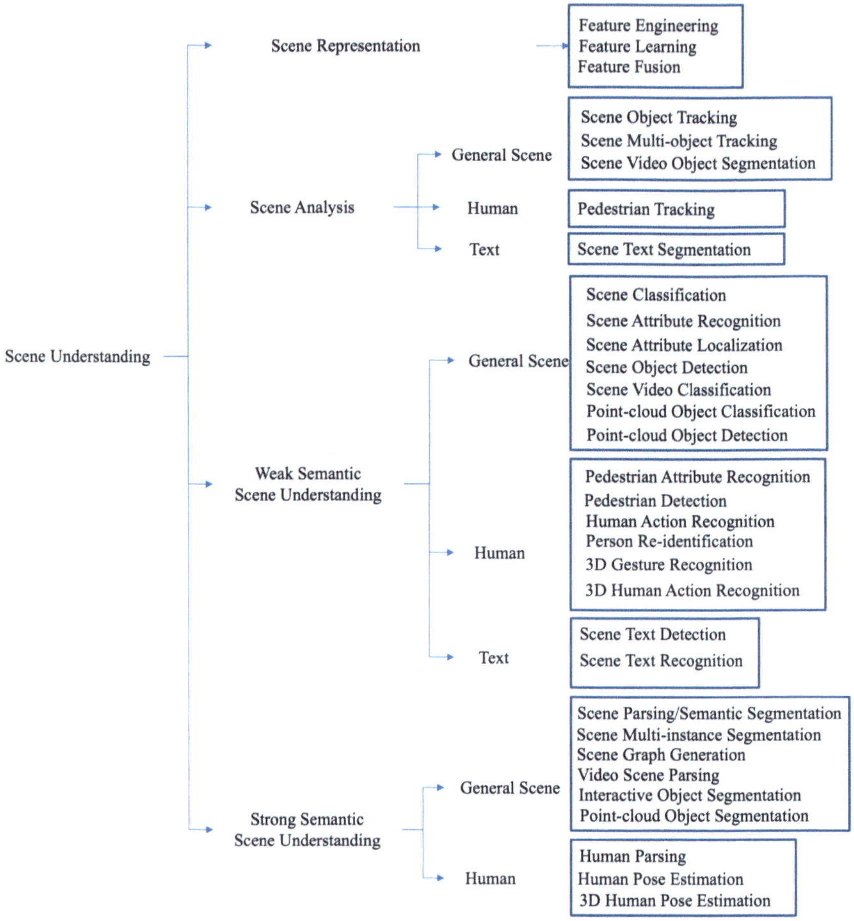

Fig. 2.4 Taxonomy of scene understanding

are not scene understanding tasks. However, the gender recognition is sometimes a sub-task of scene attribute recognition or pedestrian attribute recognition.

2.2.1 *Scene Representation*

Scene representation is the basis for scene analysis and understanding. Scene representation can be divided into three aspects: feature engineering, feature learning and feature fusion, which are described as follows.

Feature engineering refers to the manipulation of features (variables) within a dataset to enhance a machine learning model's performance. This process aims to optimize input data for modeling, facilitating improved pattern recognition and understanding of relationships by algorithms. Feature engineering seeks to automatically identify a compact set of features with minimal overlap and maximum relevance to the target variable, which is also known as feature selection. In [7], research is dedicated to examining feature selection challenges in supervised contexts. Reference [7] involves using a linear classifier to establish connections between the chosen features and class labels. Additionally, a graph regularization term is introduced to incorporate structural information into feature selection. Specifically, this term's goal is to facilitate the mapping of connected nodes with the selected features to similar class labels.

Feature learning involves acquiring the intrinsic data features that facilitate extracting valuable information for constructing classifiers or predictors [8]. Consequently, assessing a learned representation is intricately tied to its performance in downstream tasks. Deep learning is the current mainstream feature learning method, including different types of feature learning methods such as supervised learning, unsupervised learning, migration learning, reinforcement learning and transfer learning. Deep convolutional neural networks have demonstrated remarkable performance due to their inherent shift invariance, weight sharing, and ability to capture local patterns. Diverse network architectures were created to enhance model capacity, accompanied by the accumulation of increasingly large datasets. Notable networks like AlexNet [9], VGG [10], GoogLeNet [11], ResNet [12], and DenseNet [13], along with extensive datasets like ImageNet and OpenImage, have been introduced to facilitate the training of deep neural networks.

Feature fusion sometimes referred to as feature integration, is a fundamental concept within computer vision and machine learning. It revolves around the amalgamation of multiple sources of information or features to amplify the comprehension and depiction of data. The chief objective of feature fusion is to construct a more comprehensive and insightful feature representation, surpassing what can be accomplished with individual features in isolation. Reference [14] presents an efficient multi-modal authentication system based on palm-dorsa vein patterns. It begins with multi-algorithm fusion, where various vein extraction algorithms are employed to extract genuine vein patterns from the image. Data fusion is then applied to eliminate false vein patterns from the extracted data. The third fusion level combines minutiae and shape features at the feature level. In [15], the authors leverage the transformation effect between microscopic and macroscopic aspects of the general master equation to amplify the disparities in fusion probabilities among feature data units within distinct subsets. This method relies on Wootters statistical distance in probability space to identify duplicate feature data and applies a weighted median operation to fuse the identified duplicate feature data, thereby enhancing feature fusion efficiency.

2.2.2 Scene Analysis

In this section, we introduce representative methods for scene analysis. For general scenes, we introduce scene object tracking, scene multi-object tracking, scene multi-instance segmentation and scene video object segmentation methods. For two important objects in the scene, human and text, we will introduce pedestrian tracking and scene text segmentation respectively.

Scene object tracking also known as object tracking, is a computer vision method employed to trace and pinpoint particular objects as they traverse various frames or scenes in video or image sequences. Its primary objective is to consistently recognize and monitor a specific object of interest, even as it experiences alterations in position, dimensions, orientation, or temporary concealment due to occlusion. Reference [16] offers a comprehensive survey of the latest correlation-filter-based object tracking techniques, evaluating their strengths and weaknesses through both theoretical and experimental lenses. The methods are categorized based on features, space weight factors, scale factors, and expert strategies. The article further includes extensive experimentation involving 11 tracking challenges on benchmark datasets. Zhang et al. [17] argue that the update model depends on the initial ground truth, the last frame's template, and the current frame's appearance model. They represented this function as a deep network and introduced the UpdateNet specifically designed for model updating purposes.

Scene multi-object tracking (MOT) represents a sophisticated computer vision challenge wherein the simultaneous tracking of multiple objects in a dynamic scene, often within a video or image sequence, is the core task. The primary goal of MOT is to consistently monitor and trace the trajectories, identities, and interactions of multiple objects over time. Distinguishing it from single-object tracking, MOT confronts the intricacies of tracking multiple objects that can occlude each other, undergo appearance changes, and engage in intricate interactions. Reference [18] proposes a video object-centric model for MOT. The model has two key modules: an index-merge module and an object memory module which help in tracking objects consistently over time and handling occlusions respectively. The model's design focuses on tracking objects as a whole and consistently over time by consolidating slots into memory buffers and rolling past representations of memory forward. This approach provides a label-efficient alternative to the otherwise costly MOT pipelines that rely on detection and ID labels. Nguyen et al. [19] address the complex task of tracking multiple objects using a network of distributed sensors with limitations in field of view, computing power, and communication resources. They define the concept of label consistency and develop an approach to reduce label inconsistency caused by objects moving between nodes with limited fields of view.

Scene video object segmentation can be regarded as an extension of scene image segmentation in the time domain, with the main goal of segmenting the interested or specified video objects from the video and ensuring the temporal consistency of the segmentation results. RANet [20] introduces a ranking attention module to prioritize and select similarity maps between the features of the initial and current frames. However, RANet's approach of directly concatenating previous frame segmentation results within the feature space may not fully harness the potential of the previous frame's feature information. JOINT [21] introduces an integrated approach, comprising both transductive and inductive inference branches. For the transductive inference branch, it adopts a modern Transformer architecture, while employing a few-shot learner for the inductive inference branch. The authors have ingeniously devised a regularization loss to ensure the independence and complementarity of these two branches.

Pedestrian tracking which employed in computer vision and surveillance systems, is designed to monitor and trace the paths of individuals. This process encompasses the identification and tracking of people within video streams or sequences of images. The initial step involves the detection of pedestrians in a frame by a computer vision system, which discerns their presence and location based on attributes such as color, shape, or motion. Subsequently, a tracking algorithm assigns a distinct identifier to each person and persistently monitors their movements through successive frames. Reference [22] introduces an active pedestrian tracking system inspired by the human visual system. The authors employs a coarse-to-fine pedestrian detection algorithm that combines Gaussian mixture model background subtraction with histogram of oriented gradient detection, particularly tailored for small pedestrian detection. Additionally, they introduce a three-dimensional pan-tilt-zoom control model that requires no calibration and offers superior accuracy compared to other control models. To enable real-time pedestrian tracking, they utilize an active control algorithm and a tracking-learning-detection tracker. Reference [23] introduces a real-time pedestrian tracking and positioning method that utilizes multiple overhead-view cameras. The frame difference method is applied to detect pedestrians in motion as they transition between camera views.

Scene text segmentation serves the purpose of recognizing and isolating textual elements within images or video frames. Its primary objective is the separation of text from the complex background or cluttered scenes. The scene text segmentation method refines the initial detection by precisely localizing and outlining the boundaries of these text regions. Reference [24] employs text-specific features to detect candidate text boundaries based on image edges. Subsequently, the authors extract potential characters within these boundaries using a local threshold derived from surrounding image pixels. The final step involves identifying actual characters and words using a support vector regression model trained on bags-of-words representation. Reference [25] introduces a novel data-driven approach for pixel-level scene text segmentation from a single natural image. The method incorporates text region masks as auxiliary data, as acquiring labeled text region masks is typically more cost-effective and less time-consuming. The approach involves a mutually guided network, generating both polygon-level and pixel-level masks in parallel, with each branch guiding the other. Besides, training process utilizes a semi-supervised learning strategy. Furthermore, the network's output masks not only enhance text recognition but also have potential applications in image editing.

2.2.3 Weak Semantic Scene Understanding

Scene classification entails the categorization of images or scenes into predefined classes or categories determined by their content. This core objective is the automatic identification of a scene's type or category, whether it be a beach, forest, cityscape, or kitchen. This task is pivotal in computer vision and pattern recognition as it empowers machines to comprehend and interpret the visual content depicted in images and videos. Reference [26] categorizes visual scenes into two distinct types: object-based scenes, characterized by their reliance on specific objects within the scene, and layout-based scenes, which are comprehensible based on the scene's layout and surrounding context. Scene-specific objects play a vital role in understanding object-based scenes by offering semantic cues. The authors introduce a discriminative hybrid representation for visual scenes. This approach combines semantic features extracted from scene-specific objects with visual features derived from a deep CNN. Reference [27] delves into the primary challenges of remote sensing image scene classification and surveys three key categories: firstly, methods based on autoencoders for remote sensing image scene classification; secondly, approaches employing convolutional neural networks for this purpose; and thirdly, methods using generative adversarial networks. Additionly, they explore the promising avenues for future research in this field.

Scene attribute recognition encompasses the identification and annotation of diverse characteristics and features within an image or video, describing the depicted scene. In contrast to traditional scene classification, which assigns scenes to predefined categories, scene attribute recognition delves deeper, seeking to convey intricate insights about the scene's content. This approach yields a richer, more nuanced comprehension of scenes, proving particularly valuable in domains where comprehensive scene analysis is essential. In [28], the authors investigate the significance of extracting low-level features for attribute recognition in images. The research explores the impact of different low-level features, including shape, color, texture, and deep neural network features, on the learning of visual attributes. Unlike previous research, [29] focuses on discriminating scene attributes within local regions and employs these attributes as complementary features alongside object and scene features. Visual features are extracted from two separate CNN branches: one for global image features and another for local region features. Using a contextual modeling framework, these features are combined to create more distinct scene representations.

Scene attribute localization is a computer vision task that pertains to the process of detecting and pinpointing distinct attributes or features in an image that describe the content of a given scene. These attributes encompass high-level semantic qualities like "sunny", "crowded", "urban", "beach", "nighttime", or any other descriptive markers that convey insights about the scene. Reference [30] introduces a novel concept called hierarchical space tiling (HST) to discretize the vast and continuous space of scene configurations. To provide a semantic description of objects and regions within a scene, the authors enrich the HST with attributes such as nouns and adjectives. They then present a weakly supervised approach for jointly learning both the scene configurations and attributes using a dataset of natural images accompanied by descriptive text. Reference [31] involves using non-parametric Bayesian modeling to represent intricate visual scenes, primarily utilizing weakly labeled images sourced from media-sharing platforms. For these images, which only provide general annotations for objects and attributes but lack specific location or association information, their model treats objects and attributes as latent factors and explicitly accounts for their interconnections within and between superpixels.

Scene object detection encompasses the process of recognizing and pinpointing particular objects within an image that characterizes a given scene. Its significance is highlighted in a range of practical applications where comprehending and engaging with the surroundings are fundamental, rendering it a dynamic and actively researched area in computer vision. Reference [32] proposes an approach for detecting targets in remote sensing images. The method employs a two-channel system: the first channel, the object context constrained channel, leverages recurrent neural networks to examine the contextual connections between the target and the object. The second channel, the scene context constrained channel, incorporates prior scene information and Bayesian criteria to deduce the associations between the scene and the target. Reference [33] presents an efficient approach for creating a scene-specific regression model using a generic CNN-based classifier as its foundation. The method consists of two key steps: firstly, the generic CNN-based classifier is adapted to the specifics of the surveillance scene by selecting relevant kernels. Subsequently, context-awareness is integrated into the regression model to enhance the precision of location predictions.

Scene video classification is a computer vision task which videos are systematically sorted into categories based on the scenes they portray. Essentially, it revolves around the automated assignment of a descriptive label or category to a video, encapsulating its central theme or subject matter. This technology finds utility across a spectrum of applications. Reference [34] involves employing redundant spatial grouping of numerous spatio-temporal patches, referred to as scenelets, to depict a dynamic scene. Each scenelet is linked to a category-dependent model designed to represent the likelihood of a specific scene category. These scenelet models, related to a particular scene category, are collectively learned to capture spatial interactions and redundancies among them. Reference [35] introduces a two-stage framework for Satellite Video Scene Classification (SVSC), a crucial aspect of satellite video analysis. SVSC aids in detecting objects and dynamic events within the videos. The first stage focuses on extracting spatial features by selecting representative frames through blur detection and spatial activity assessment. The second stage aims to create motion representations for the videos. It achieves this by utilizing the second temporal principal component from principal component analysis for moving targets. These two-stage features capture the spatial and temporal characteristics of satellite scenes, which are subsequently combined to enhance SVSC.

Point-cloud object classification stands as a foundational undertaking in the realm of computer vision and the analysis of 3D data. Its core mission is the recognition and grouping of items or entities nestled within a 3D point cloud. Point clouds, serving as intricate data structures, comprise an abundance of spatial points in three-dimensional space. These data sets are typically derived from 3D scanners, or LiDAR devices, and so on. Reference [36] addresses the intricate task of accurately classifying ground objects within complex urban environments using airborne laser scanning point clouds. The framework's approach involves breaking down the point cloud into hierarchical clusters of varying sizes, guided by a natural exponential function threshold. Leveraging these hierarchical point cluster relationships, the framework combines latent Dirichlet allocation and sparse coding techniques to extract and encode the shape features of these multilevel point clusters. Reference [37] introduces a robust and efficient method for classifying point clouds. It enhances feature accuracy and reduces computational complexity by incorporating point cloud supervoxels and their locally convex connected patches into a random forest classifier. To account for the diversity in point cloud feature descriptions, the approach categorizes features into three groups: point-based, eigen-based, and grid-based. Consequently, it develops three distinct feature calculation strategies tailored to enhance the reliability of these features.

Point-cloud object detection is a key task in computer vision and machine learning, dedicated to the recognition and precise positioning of objects in 3D point clouds. These point clouds are data representations of the three-dimensional environment, consisting of spatial points, each associated with specific 3D coordinates, and potentially other attributes like color or intensity. This technology plays a pivotal role in a variety of applications. Reference [38] introduces a novel object detection method designed to enhance robustness against variations in point cloud resolution. The approach includes two key components: firstly, during the training phase, random discarding of point cloud layers is employed to diversify the input data processed by the network. Secondly, obstacles are represented as Gaussian functions, consolidating multiple parameters into a single representation. Reference [39] introduces an 3D object detection system, aiming to grasp both geometric and semantic aspects of point cloud features. This method performs localized feature extraction using the gateway attention pooling module, enhancing the discriminative nature of point cloud characteristics. Leveraging high-quality 3D proposals generated by attention-based backbone networks, they implement a RoI multi-pooling head to adaptively acquire features from sparse points of interest within these proposals.

Pedestrian attribute recognition is a task focused on generating structured descriptions of pedestrians, such as color, clothes, age, etc. This task is of significant importance in surveillance applications. Pedestrian attribute recognition is a multi-label classification task for detecting pedestrian attributes. However, accurate recognition of pedestrian attributes can be challenging due to various factors like diverse illumination conditions, partial body occlusion, and limitations in image resolution. These challenges make it essential to develop robust algorithms and models to improve recognition results under such conditions. An et al. [40] observe that the existing methods address this problem by approaching it as a multi-label classification task without taking into account the hierarchical relationships among the attributes. Hence, they propose a novel Hierarchical Reasoning Network (HR-Net), which can hierarchically predict the attributes at different abstraction levels in different stages of the network. Fan et al. [41] propose a relationship framework for comprehensively describing and utilizing relations among attributes named Correlation Graph Convolutional Network (CGCN). By considering various types of relations in the same dimension and implementing complex transfers of relations in a Graph Convolutional Network (GCN) manner, this framework provides a more accurate and context-aware recognition of pedestrian attributes, even in challenging conditions. In this framework, feature vectors are employed to associate attributes with image features and create various relation matrices using self-attention mechanisms among different feature vectors. Then, they conduct multi-layer transfer of attribute relations by means of graph convolution, realizing complex utilization of attribute relations.

Pedestrian detection is a challenging problem in the field of computer vision, and it has numerous practical applications in the real world. The majority of existing approaches utilize a holistic detection strategy, which assumes that pedestrians are entirely visible when trained with full-body annotations. While promising results have been achieved for standard pedestrians, the performance on heavily occluded pedestrians is still far from satisfactory. Toward this, Xie et al. [42] present an approach for detecting occluded pedestrians with several key contributions. First, they introduce a novel mask-guided attention network designed to seamlessly integrate into popular pedestrian detection pipelines. Second, they propose a method for occlusion-sensitive hard example mining and an occlusion-sensitive loss function. These techniques identify challenging examples based on the level of occlusion and assign greater importance to detection errors in highly occluded pedestrian cases. Third, they empirically validate that weak box-based segmentation annotations offer a reasonable approximation to their dense pixel-wise counterparts. Pedestrian detection, which relies on deep convolutional neural networks, has made significant progress. Further, Liu et al. [43] propose to perform contrastive learning for pedestrian detection task. Concretely, they employ the contrastive learning to guide the feature learning. This can minimize the semantic distance between pedestrians with different appearances in the learned feature space, can then eliminate the appearance diversities. Besides, to enhance the efficiency and efficacy of contrastive learning, we create an exemplar dictionary containing representative pedestrian appearances as prior knowledge. This dictionary assists in generating effective contrastive training pairs, guiding the contrastive learning process.

Human action recognition serves as a pivotal component in numerous human-computer interaction applications. The primary objective of Human Action Recognition (HAR) is to categorize specific human actions within videos into predefined action categories. Examples of these categories may include actions like "take off a hat", "take off glasses", "clapping", "hand waving", and so on. Liu et al. [44] introduce the initial adversarial attack designed for skeleton-based action recognition using Graph Convolutional Networks (GCNs), termed constrained iterative attack for skeleton actions (CIASA). CIASA accomplishes this feat by simultaneously adhering to multiple physical constraints, utilizing spatial skeleton realignments for the perturbed skeletons, and applying regularization techniques to the adversarial skeletons through generative networks. Most current research efforts concentrate on enhancing model generalization by incorporating multiple homogeneous modalities, such as RGB images, human poses, and optical flows. Based on the dataset and inspired by the idea that different modalities can exhibit temporal consistency and spatial complementarity, Guo et al. [45] design a model, named Bi-directional Co-temporal and Cross-spatial Attention Fusion Model (B2C-AFM). B2C-AFM is characterized by the asynchronous fusion approach, which combines multi-modal features along both temporal and spatial dimensions.

Person re-identification is a cross-camera instance retrieval problem that focuses on the task of identifying individuals across multiple non-overlapping cameras. This problem has garnered significant research attention, with most existing works primarily concentrating on supervised learning approaches. In order to solve the burden of massive annotation requirement, Wang et al. [46] introduce a person re-identification model from videos with weak supervision. Towards this goal, they propose a multiple instance attention learning framework for person re-identification using weak supervision. Specifically, they initially cast the video person re-identification task into a multiple-instance learning problem, where person images within a video are grouped into a "bag". Then, they introduce a co-person attention mechanism that explores the similarity correlations between videos that share common person identities. After this, optimizing deep re-ID models and acquiring discriminative person representations without annotations in the training data remains a challenging task. Hence, Yang et al. [47] introduce a novel approach to solve the unsupervised person re-ID by leveraging virtual and real data. Concretely, this model includes two components: virtual person generation and training of the deep re-ID model. Specially for virtual person generation, we train a model for generating virtual people and another model for transferring camera styles. We accomplish this using unlabeled real data, enabling us to create virtual individuals with various poses and camera styles.

3D gesture recognition is a technology that captures and interprets human gestures in three-dimensional space to enable interaction with computers or other devices. As artificial intelligence technology advances, human 2D pose estimation has become a powerful tool for understanding human actions. However, it's important to note that 2D pose estimation only captures the planar aspects of the actions, which can introduce spatial ambiguity. Especially for traffic police gestures, the collected traffic police scales change all the time, in addition, the frequency and amplitude of actions of different traffic police are different. Hence, Fu et al. [48] propose a fully adaptive spatial-temporal graph convolution network (FA-STGCN) which can recognize the actions with different amplitude and frequencies. The adaptive spatial graph network is capable of uncovering the latent connections between joints in the traffic police's body under various gestures. This helps reduce the influence of amplitude variations on action recognition. The adaptive temporal graph network consists of two components: the global temporal module and the local temporal module. The global temporal module is responsible for capturing the coarse-grained features related to the speed of traffic police gestures. It then utilizes these coarse-grained features to guide the local temporal module in adaptively learning the finer-grained temporal features of traffic police actions. Liu et al. [49] describe a 3D skeleton as a point within the special orthogonal group space. This representation is specifically designed to model the 3D geometric relationships between different body parts. Consequently, a skeletal sequence representing a gesture can be characterized as a trajectory on a Riemannian manifold. Building on this concept, they introduce a generalization of the transported square-root vector field to derive a time-warping invariant metric for comparing these trajectories, thereby enabling the identification of these gestures. Further, Gao et al. [1] propose to apply 3D hand pose estimation, data fusion and deep neural network to improve the recognition accuracy of dynamic hand gestures. They employ the Openpose to obtain a fast 3D hand pose estimation method. Then, the proposed method combines the RGB, depth and 3D skeleton data of hand gestures by a weighted sum fusion method. Finally, a framework combining 3DCNN and ConvLSTM is employed for the identification and classification of the dynamic hand gesture data.

3D human action recognition is a technology that identifies and classifies human actions or behaviors in three-dimensional space using data captured from various sensors. With the rapid advancement of motion capture technology, the field of 3D human action modeling and recognition has garnered increasing interest due to its significance in a wide range of applications. 3D human action data is frequently represented in the form of RGB-D or depth values of the human body, or as 3D human skeletons that depict the three-dimensional positions of human body joints. Wei et al. [50] suggest that latent states possess composite attributes, and they introduce a novel Composite Latent Structure (CLS) model to represent and recognize 3D human actions based on skeleton sequences. A human action is modeled using a hierarchical graph, which portrays the action sequence as a series of sequential atomic actions. Each atomic action is represented as a composite latent state, comprising both a latent semantic attribute and a latent geometric attribute. Finally, they introduce a discriminative EM-like algorithm to learn both the model parameters and the composite latent structures associated with human actions. Li et al. [51] propose a symbiotic model to handle 3D skeleton-based action recognition and motion prediction jointly, and propose two scales of graphs to explicitly capture relations among body-joints and body-parts. For the backbone of the model, they introduce a multi-branch multiscale graph convolution network, which is utilized to extract both spatial and temporal features. These multiscale graph convolution networks rely on joint-scale and part-scale graphs as their foundation. The joint-scale graphs consist of actional graphs, which capture relations related to actions, and structural graphs, which capture physical constraints among the body-joints and body-parts.

Scene text detection plays a pivotal role in scene text reading systems. The primary challenges in this task revolve around handling significantly diverse text sizes and aspect ratios, accommodating arbitrary orientations, and identifying various text shapes. Xu et al. [52] introduce a novel text detection method named TextField which is designed for the detection of irregular scene texts. In this approach, they train a fully convolutional neural network to learn a direction field that points away from the nearest text boundary for each text point. This direction field is represented as a 2D vector image and encodes both binary text masks and direction information.

This information is invaluable for separating adjacent text instances, which is a challenging task for traditional segmentation-based methods. Taking a hybrid approach, Bi et al. [53] investigate hierarchical text instance-level and component-level representations for the detection of arbitrarily-shaped scene text. Based on this, they propose a novel Hierarchical Graph Reasoning Network (HGR-Net), which consists of a Text Feature Extraction Network (TFEN) and a Text Relation Learner Network (TRLN). In TFEN, the model adaptively learns text candidates at multiple granularities using shared convolutional feature maps. In TRLN, an inter-text graph is built to explore global contextual information with an awareness of the positions of text instances. Additionally, an intra-text graph is created to estimate geometric attributes, facilitating the establishment of component-level linkages.

Scene text recognition is a challenging task within the field of computer vision, primarily due to the wide variety of text styles and the complexity of image backgrounds. Over the past few decades, many text rectification and recognition methods have been developed to address these challenges. Wu et al. [54] introduce a Two-level Rectification Attention Network (TRAN) for the rectification and recognition of text. This network is comprised of two components: a TwO-level Rectification Network (TORN) and an Attention-Based Recognition Network (ABRN). The TORN is responsible for the initial rectification of text at the geometric level, followed by pixel-level adjustments. This process effectively removes geometric constraints and enhances text clarity. The ABRN, on the other hand, is responsible for recognizing text within the rectified images. Li et al. [55] introduce a Character-Aware Sampling and Rectification (CASR) module, which is designed to rectify irregular text instances based on the location and orientation information of each individual character within the text. Specifically, CASR treats each character as a fundamental unit and predicts character-level attributes for the purposes of sampling and rectification. This module utilizes detailed character information to improve the rectification of text lines and incorporates character-level supervision during the training process.

2.2.4 Strong Semantic Scene Understanding

Scene parsing and semantic segmentation are the tasks of assigning predefined categories of things and stuff labels to all pixels in an image. A fundamental requirement for this task is the acquisition of a robust representation for each pixel. Shi et al. [56] present a novel approach, Hierarchical Parsing Net (HPN), specifically designed for scene parsing. In contrast to conventional methodologies that classify individual objects in isolation, HPN strategically utilizes global scene semantic information and inter-object context to enhance scene parsing accuracy. On one hand, HPN employs the global scene category to enforce semantic consistency between the scene and each individual object. Concurrently, the model comprehensively models the context among all objects to avoid incompatible object predictions. However, this network discovers desired features or representations only from the given image (content), ignoring more generic knowledge contained in the dataset. To overcome this deficiency, Luo et al. [57] attempt to investigate the integration of meaningful supportive knowledge into scene parsing. This knowledge encompasses general visual concepts, representing generic object and stuff categories, along with their interrelations derived from the entire dataset. To this end, they introduce a novel Supportive Knowledge Mining Module (referred to as SKMM) and a Knowledge Augmentation Operator (KAO), which can be easily plugged into modern scene parsing networks.

Scene multi-instance segmentation is also known as instance segmentation. The instance segmentation is a computer vision challenge centered on the precise identification and segmentation of individual object instances within a scene, commonly found in images or video frames. This task seamlessly integrates two core objectives: object detection and instance segmentation. It entails not just detecting objects but also the precise delineation of each instance by assigning a distinct mask, even when objects share the same category, enabling differentiation among them. In [58], the authors introduce the concept of "instance categories." This assigns categories to pixels within an instance based on their location. This novel perspective leads to the development of "Segmenting Objects by LOcations" (SOLO), a straightforward and efficient framework for instance segmentation with impressive performance. Reference [59] introduces Sem2Ins, a novel real-time model that bridges the gap between semantic segmentation and instance segmentation. The authors achieve this by utilizing conditional Generative Adversarial Networks (cGANs), along with deep supervision and a weighted fusion layer, to effectively generate instance boundaries based on a semantic segmentation.

Scene graph generation is a fundamental process in the realms of computer vision and natural language processing. This task revolves around the meticulous crafting of a structured depiction of objects, their characteristics, and the intricate connections they share within an image or a scene. It constitutes an indispensable stride in comprehending visual scenes, which can be applied to image captioning, action recognition, etc. Reference [60] introduces an end-to-end model called Relation Transformer. This method proposes an encoder-decoder architecture where the encoder focuses on visual context, and the decoder generates a set of triplets using various attention mechanisms with linked subject and object queries. The method directly predicts sparse scene graphs based solely on visual cues, without the need for entity combination or exhaustive predicate labeling. Furthermore, [61] concludes that the mean field variational Bayesian framework tends to overlook constrained optimization models. The authors introduce a novel approach to structured learning, emphasizing explicit constrained variational inference. Unlike the common message-passing approach, we employ entropic mirror descent, a generic optimization method, to address the constrained variational inference task.

Video scene parsing is a longstanding and challenging task in the field of computer vision. Its objective is to assign predefined semantic labels to pixels in all frames of a given video. Compared with image semantic segmentation, this task pays more attention on investigating how to leverage temporal information to achieve higher predictive accuracy. Miao et al. [62] introduce a novel dataset with the primary objective of advancing the scene parsing task from static images to dynamic videos. This dataset denoted as Video Scene Parsing in the Wild (VSPW) covers a wide range of real-world scenarios and categories. Utilizing the VSPW dataset as a foundation, they propose a novel Temporal Context Blending (TCB) network. This network is engineered to effectively leverage long-range contextual information from past frames, significantly enhancing the segmentation process for the current frame. Kim et al. [63] introduce VPSNet++, an advanced video panoptic segmentation network, which simultaneously performs classification, detection, segmentation, and tracking of all identities in videos. Specifically, VPSNet++ builds upon a top-down panoptic segmentation network by adding pixel-level feature fusion head and object-level association head.

Interactive object segmentation is a method that enables human annotators to rapidly extract the object of interest by providing user inputs such as bounding boxes or clicks. Concretely, most interactive object segmentation methods are aimed at reducing human effort required for pixel-level annotations by utilizing guidance such as bounding boxes, clicks, or contours. GrabCut [64] is a pioneering work in the field of interactive segmentation, utilizing bounding boxes to guide the segmentation process. Then, Zhang et al. [65] explore how to harvest precise object segmentation masks while minimizing the human interaction cost, and propose an Inside-Outside Guidance (IOG) approach. IOG utilizes two points outside the region of interest effectively which can aid in removing distractions from other objects or background. Utilizing an inside point is instrumental in eliminating unrelated regions within the bounding box. Additionly, IOG seamlessly accommodates additional click annotations for further refinement and correction. Xu et al. [66] present an interactive framework for image segmentation and visualization, specifically designed to identify, inspect, and edit minuscule objects (a few pixels wide) in large multimegapixel high-dynamic-range (HDR) images. We developed an interactive toolkit that unifies model inference, HDR image visualization, segmentation mask inspection and editing into a single graphical user interface.

Point-cloud object segmentation is a process in computer vision and 3D data analysis that involves dividing a point cloud into distinct segments, each corresponding to different objects or parts of objects. The success achieved by image-based methods has been effectively extended from the 2D domain to 3D point cloud semantic segmentation. Nie et al. [67] explore efficient and effective strategies for integrating features at diverse scales and stages within a point cloud segmentation network. They open up the conventional encoder-decoder architecture and introduce novel scale pyramid architectures. These architectures allow information to flow more freely and systematically, both laterally and upward/downward in scale. Shuai et al. [68] analyze the geometric distortion between the point cloud and the pseudo-image, including truncation, dislocation, and hole. Based on this, they propose the Geometry-injected Image-based point cloud semantic segmentation Network (GINet). The GINet rectifies the geometric distortion while taking advantage of the successful 2D semantic segmentation methods.

Human parsing is considered a fundamental task in computer vision as it involves accurately segmenting the various semantic human body parts at the pixel level. This task has garnered significant attention in recent years due to its relevance in a wide range of human-centric applications. Zhang et al. [69] design a suitable mechanism to collect spatial clues aligning with different human parts. Concretely, they use a Gather-Excite (GE) operation, which replaces the AvgPool-Upsample operation in a pyramidical structure, to accurately capture relevant information from human parts of different scales. The majority of state-of-the-art instance-level human parsing models employ two-stage anchor-based detectors. However, these models face challenges related to heuristic anchor box design and a lack of detailed pixel-level analysis. Toward this, Song et al. [70] have developed an instance-level human parsing network that is anchor-free and capable of solving pixel-level tasks. The proposed method consists of two simple sub-networks: an anchor-free detection head for bounding box predictions and an edge-guided parsing head for human segmentation. Specially, by incorporating the part-aware boundary clue, the edge-guided parsing head can differentiate between adjacent human parts, even in the presence of up to 58 parts within a single human instance or overlapping instances.

Human pose estimation is a computer vision task that involves determining the locations and orientations of key human body joints or keypoints from images or videos. This task is fundamental in understanding human behavior, gesture recognition, action recognition, and a wide range of applications, including robotics, sports analytics, healthcare, and virtual reality. Ke et al. [71] introduce a DetPoseNet, a comprehensive end-to-end framework for multi-human detection and pose estimation, organized within a unified three-stage network. This method consists of a coarse-pose proposal extraction sub-net, a coarse-pose based proposal filtering module, and a multi-scale pose refinement sub-net. Specially, this method is highly adaptable and can seamlessly integrate with a variety of existing top-down pose estimation methods, with the pose refinement sub-network playing a pivotal role in our approach. Precision in whole-body multi-person pose estimation and tracking is a fundamental in the field of computer vision. Fang et al. [72] introduce AlphaPose, a real-time system designed for simultaneous and precise whole-body pose estimation and tracking. Toward this, they propose several new techniques, including Symmetric Integral Keypoint Regression (SIKR) for fast and fine localization, Parametric Pose Non-Maximum-Suppression (P-NMS) for eliminating redundant human detections and Pose Aware Identity Embedding for jointly pose estimation and tracking. This method can accurate localizate the whole-body keypoints and simultaneous carry out human tracking, even when provided with imprecise bounding boxes.

3D Human Pose Estimation is a computer vision task that involves determining the precise positions and orientations of a person's joints in three-dimensional space from visual data, such as images or videos. The estimation of a 3D human pose from a 2D image has emerged as a pivotal component in the visualization and analysis of pose for upcoming next-generation applications. However, the task of estimating 3D poses from a single image has historically been a formidable challenge, primarily due to the several roadblocks such as diverse posture form, clothing, colors, and self-occlusion. Lee et al. [73] propose a perceptual Pose SIMilarity (PSIM) metric, by assuming that human perception (HP) is highly adapted to extracting structural information from a given signal. Second, they introduce a perceptually robust 3D pose estimation framework known as Temporal Propagating Long Short-Term Memory networks (TP-LSTMs). Toward this, they conduct an analysis of information-theory-based spatio-temporal posture correlations, which encompass joint interdependency, temporal consistency, and HP. 3D human pose estimation methods frequently exhibit reduced generalization performance when applied to new datasets. This issue primarily stems from the limited diversity of 2D-3D pose pairs in the training data. To address this challenge, Zhang et al. [74] introduce PoseAug, an innovative auto-augmentation framework designed to learn and apply augmentations to the available training poses. This augmentation process significantly enhances the diversity of the training data and, as a result, bolsters the generalization capability of the trained 2D-3D pose estimator.

2.3 Three Key Tasks of This Book

Scene understanding includes a variety of visual tasks. But in this book, in order to present advantage of graph neural network on different level of scene understanding, we mainly focus on three typical tasks. The first one, video object segmentation, is a kind of low-level task in scene analysis, the other two, scene parsing and human parsing, are high-level tasks in strong semantic scene understanding. And the human parsing is a kind of much finer-grain scene understanding task with semantic labels having much stronger structure dependencies than scene parsing.

This section will provide a detailed introduction to the three tasks. Video object segmentation is a computer vision task that aims to segment and track specific objects throughout the entire video, providing a frame-by-frame prediction result of their boundaries. This task is crucial for various applications in scene understanding, including surveillance, autonomous vehicles, and augmented reality. Scene parsing is a key technology to realize semantic segmentation and recognition of image. It requires the algorithm model to realize pixel-level scene object segmentation and recognition, which is not only a multi-label problem, but also needs to match or locate labels to specific image pixels. Human parsing, as a fine-grained semantic segmentation task, focuses on accurately identifying the various parts of the human body. It plays a crucial role in numerous downstream applications, such as crowd trajectory prediction, action recognition, virtual try-on, and virtual reality.

2.3.1 *Scene Parsing*

Scene parsing is a complicated task, and it is also one of the emphases of future computer vision research. It requires the algorithm model to realize pixel-level scene object recognition and segmentation, which is not only a multi-label problem, but also the need to match or locate labels to specific image pixels, which is often called image semantic segmentation. This requires more intensive annotation of scene images that contain a large number of objects. For example, given a visual scene of an indoor living room, it is necessary to label not only objects like sofas, tables, and televisions, and to divide background areas like floors, walls, and ceilings, but also more granular components like the seat of a chair or the handle of a cup. So that the algorithm model can learn fine scene features from the original data, and realize pixel level recognition and localization.

One of the common design concepts of modern intelligent systems is the ability to respond to different situations depending on the field environment without human intervention. Therefore, it is important for such systems to be able to accurately sense their location or environment, which helps them to retrieve events that have already occurred and predict future changes, and then decide what to do next. For example, a mobile robot is such an intelligent system, it may pass through different places in the process of reaching its destination. Hence, it need to adjust its actions in real time to adapt to the changing environment along the way. The behavioral decisions of mobile robots depend on past, present and future environmental information. The scene parsing can not only recognize the target object in the environment, but also provide the specific location information and semantic information of each target object. Based on this, the intelligent system can generate the basic description of the target, event and environment in the scene, which helps the intelligent system to perceive and understand the surrounding environment. As a result, scene parsing is widely used in applications such as intelligent human-computer interaction, mobile robotics, intelligent video surveillance, and autonomous driving, and it is also regarded as a prerequisite or prior knowledge for other computer vision tasks such as image retrieval and image description.

Scene parsing is one of the most challenging and important problems in computer vision in recent years. The scene can be a natural area in which humans can move, or it can be an artificial area. A scene is generally a combination of objects (target information), places (environment information), and contexts (event information). For the current computer system, it is difficult to directly learn the connotation pattern of the whole scene. Therefore, it is necessary to establish a scene representation model in the computer by means of feature extraction and feature learning to indirectly express the scene information. The basic goal of scene parsing is based on the scene representation model. Usually, the algorithm model assigns semantic labels to a given scene image, and these semantic labels are artificially defined. Macro types of scenes generally include natural scenes, artificial scenes, and the latter includes indoor scenes, outdoor scenes and so on. Each type of scene can be divided into a variety of specific categories, for example, natural scenes include beaches, rivers, grasslands, etc., indoor scenes include bookstores, restaurants, waiting rooms, etc., and outdoor scenes include streets, ports, highways, etc. Compared with the general image classification studied in the field of computer vision, scene parsing requires not only the recognition of objects, but also the perception of the distribution of objects in the environment and the discrimination of environmental attributes. Therefore, scene parsing is more focused on the analysis, discrimination and inference of high-level image features.

Through the above discussion and analysis of scene parsing, there are still many challenges in achieving more general scene parsing due to the high complexity of scene images. In the past ten years, with the rapid popularization of high-performance computing devices, deep learning methods have become the mainstream and received a lot of research and development. In the field of computer vision and pattern recognition, deep learning models have also made significant progress, and a variety of artificial intelligence product technologies formed based on this method are also rapidly applied in transportation, urban construction, daily life and office. In these application environments, scene parsing algorithms based on deep learning have made certain breakthroughs in performance, while methods based on classical feature engineering have received less and less attention. However, with the deepening of research on deep learning methods, it has gradually entered a bottleneck. Faced with increasingly complex scene categories, attribute types and semantic relations, it is urgent to combine more basic theories and methods to find new breakthrough points. However, this research finds that feature fusion method still has its advantages: by combining original features to remove information redundancy and irrelevance, it can tolerate some defects in feature selection and extraction, provide complementary information, improve the accuracy of the overall decision process, and simplify the difficulty of feature selection in classification problems. In fact, feature fusion is still one of the most attractive research hot-spots, and many researchers are studying methods to achieve feature fusion in combination with other models, such as statistical models, sparse models, or depth models.

2.3.2 *Human Parsing*

Human parsing, which aims to precisely segment the different semantic human body parts on the pixel level, is regarded as one of the fundamental tasks in computer vision. As it finds a wide spectrum of human-centric applications, human parsing has attracted a lot of attention in recent years. Human parsing is a more challenging task compared to traditional image segmentation because it requires distinguishing between subtle variations in body parts, poses, clothing styles, and interactions. The resulting segmentation map provides a comprehensive description of the person in the image, enabling various applications such as person re-identification [75], human behavior analysis [76], clothing style recognition and retrieval [77], and clothing category classification [78]. In the video surveillance, human parsing can analyze frames of video images, and identify individuals based on the attributes of their body parts. This information can be utilized for enhancing security measures and aiding investigations. In the clothing recognition, human parsing technology can be employed to identify the location and size of clothing within an image. This enables functionalities like fashion search and replacement, and can offer users an intelligent clothing recommendation experience and optimize consumers' shopping experience. For effective human-computer interaction, machines need to comprehend human actions and behaviors. Human parsing algorithms can assist machines in recognizing human limbs, which has valuable implications for analyzing human movements and behaviors, thus improving interactions between humans and computers.

The rapid advancement of deep learning techniques [79, 80] and the availability of vast amounts of image data [81, 82] have led to significant progress in the field of visual understanding. Notably, achievements such as surpassing human-level accuracy in face recognition [83] and real-time human pose estimation on mobile devices [84] highlight the continuous development in this area. However, the exploration of more intricate tasks, such as fine-grained human parsing and multi-label clothing attribute classification, is still in its early stages. Human parsing is a task in computer vision that focuses on dissecting an entire human body within an image into distinct components such as body parts and different clothing regions. These parts can include elements like the head, neck, coat, pants, and more. Similar to object semantic segmentation, human parsing is a pixel-level classification challenge where the goal is to assign a semantic label to every individual pixel in the image. This means that the objective is to assign a semantic label to every pixel in the image, accurately categorizing each pixel based on its corresponding body part or clothing type. Thus we can achieve a detailed understanding of the human body in the image.

Compared with other semantic segmentation, human parsing benefits from inherent hierarchical and topological relationships that exist between different body parts and clothing items. This structure provides a rich source of information that can be leveraged to improve segmentation accuracy and feature extraction. Specially, the human body can be naturally divided into different levels of semantic information. For instance, body parts can be categorized into higher-level groups like "upper body" and "lower body", and further into more specific segments like "arms", "torso", "pants", etc. This hierarchical information can guide the segmentation process, helping to ensure that the predictions align with the anatomical structure of the human body. Further, the human body exhibits strong topological relationships between its different parts. For example, limbs are connected to the torso, and clothing items are worn on specific body regions. These relationships are consistent across different individuals and scenarios. This inherent structure provides valuable constraints that can aid in accurate segmentation and recognition. In contrast, scene segmentation may not have such clear hierarchical and topological relationships between objects. Objects in a scene can be diverse and unrelated, making it challenging to define a consistent hierarchy or capture strong constraints between them. By taking advantage of the stable hierarchical and topological structure of the human body, human parsing algorithms can achieve more accurate and contextually meaningful results. These insights allow researchers to design human parsing models that effectively leverage the inherent structure of the human body. On the other hand, unlike medical image segmentation, where the background is often a relatively controlled and consistent environment, human parsing is applied to a wide range of scenarios. This can include diverse backgrounds such as streets, indoor spaces, scenic locations, and more. The presence of dynamic and varied backgrounds introduces challenges in distinguishing between the body and the background, especially when the background may have similar colors, textures, or patterns to the body or clothing. The varying backgrounds in human parsing can indeed introduce interference to the segmentation process. The model needs to be robust enough to handle these variations and still accurately identify and segment the human body parts. This requires the segmentation algorithm to effectively differentiate between the foreground (human body) and the background, even when the background is complex and diverse. To address these challenges, human parsing algorithms often incorporate techniques such as semantic context modeling, attention mechanisms, and data augmentation to improve background distinction and segment the human body accurately. Additionally, advancements in deep learning and the availability of large-scale annotated dataset have contributed to the development of more robust and effective segmentation models for human body analysis, even in the presence of diverse and dynamic backgrounds.

Human parsing requires an abundance of pixel-level annotations for fully supervised, and accurate segmentation of different body parts and clothing areas requires detailed annotations for each pixel. The pixel-level annotations are a time-consuming and labor-intensive process. The complexity of the human body structure and the need for precise segmentation make human parsing even more challenging. To address this issue, researchers often explore weakly supervised learning and semi-supervised learning to reduce the annotation burden while maintaining or improving model performance. Instead of annotating every pixel, weakly supervised learning methods use less precise annotations like bounding boxes or image-level labels to guide the learning process. This reduces annotation effort while still allowing the model to learn useful features. The challenge of weakly supervised human parsing lies in developing techniques that can effectively leverage this limited or noisy supervision to generate accurate pixel-level segmentation of different body parts. For weakly supervised human parsing, Zhao et al. [85] propose a part evolution framework to learn reliable predictions from weak pose annotations, which are much easier to collect. Semi-supervised methods utilize a combination of labeled and unlabeled data. This can help leverage the available annotations more effectively and potentially reduce the need for extensive pixel-wise annotations. Unlike fully supervised human parsing, which requires pixel-level annotations for training, and weakly supervised human parsing, which uses weaker forms of annotation, semi-supervised approaches aim to leverage a smaller amount of labeled data along with a larger amount of unlabeled data to improve segmentation performance. In semi-supervised human parsing, the labeled data typically consists of images with pixel-level annotations. The unlabeled data, on the other hand, only includes images without such detailed annotations. The challenge is to develop techniques that can effectively utilize the information from both labeled and unlabeled data to produce accurate segmentation. Fang et al. [86] propose a weakly and semi-supervised human parsing, which transfers the pixel-level parsing annotations from the fully supervised sample to the weakly supervised sample based on their similar poses. Furthermore, Liu et al. [87] begin to explore the effectiveness of this hierarchical structure in semi-supervised human parsing. They develop a noise-tolerant method to handle the noisy pseudo labels, and then parsing the human body in a coarse to fine manner could better assist the training of the semi-supervised model. In summary, the difficulty of weakly supervised human parsing and semi-supervised human parsing lies in how to generate accurate pseudo labels, and how to design a good network structure that can extract accurate information from noisy labels to achieve high-performance segmentation.

2.3.3 Video Object Segmentation

Video object segmentation [88–91] is one of the important methods to achieve intelligent scene understanding. Video object segmentation can be regarded as an extension of image segmentation in the time domain, with the main goal of segmenting the interested or specified video objects from the video and ensuring the temporal consistency of the segmentation results. Video object segmentation is widely applied in many fields, such as video editing, video encoding, action recognition, video surveillance, and autonomous driving.

On the one hand, the study of video object segmentation methods is important for massive video processing, content understanding and meeting the growing demand for video processing. On the other hand, while benefiting from advances in other computer vision fields, the video object segmentation methods also promote the research of computer vision methods such as object detection. With the in-depth research of deep learning methods in the field of video object segmentation, video object segmentation methods shift from manually designing features to automatically extracting features. Deeper relationships in the video such as temporal and spatial correspondences are mined, resulting in a higher-level video representation, which is generalized in the fields of object detection, object tracking, and other computer vision fields. In summary, the study of video object segmentation methods holds significant application value and crucial scientific theoretical significance.

According to different classification criteria, video object segmentation can be divided into different categories. Specifically, it is divided into single object segmentation and multi-object segmentation according to the number of segmentation objects. Video single-object segmentation involves segmenting a single specified object or a single object with obvious motion in the video. The resulting segmentation is a binarized mask, distinctly representing the segmented object (also known as the foreground) and the background. In most scenarios, the foreground in video single-object segmentation refers to an object that is the target to be segmented. In some videos, the foreground contains several objects, such as "man pushing a stroller", where both the man and the stroller are considered as the foreground, and similarly "man riding a horse", "carriages in common motion", because they interact with each other or have common movements. Early video object segmentation algorithms are unable to distinguish between these objects. After further development, video object segmentation methods have matured significantly, and researchers have made substantial progress. Particularly, the focus has shifted towards the more challenging task of video multi-object segmentation. Video multi-object segmentation involves designating multiple objects in the first or middle frame as reference points for segmenting subsequent video frames. This technique boasts a broader range of application scenarios in practical life. However, video multi-object segmentation also presents greater segmentation difficulties compared to video single-object segmentation.

Video Object Segmentation (VOS) can be categorized into unsupervised VOS, semi-supervised VOS, interactive VOS, and multimodal VOS based on given input supervisory information, among which, multimodal video object segmentation is an emerging video object segmentation task. In recent years, as the performance of multimodal tasks such as text-image retrieval continues to improve and multimodal-related methods make breakthroughs, multimodal video object segmentation has gradually attracted the attention of many researchers. The definitions of these four types of segmentation tasks are described below.

Unsupervised video object segmentation. Unsupervised video object segmentation is the earliest video object segmentation task. Unsupervised video object segmentation has evolved from the task of moving object segmentation and has gradually become a current research hotspot. When segmenting a video sequence, the task does not provide any supervisory information, and the algorithm automatically segments objects with obvious motion in the video, which is known as Zero-shot Video Object Segmentation (ZVOS). The ZVOS methods [92–96] automatically extract the main objects in videos without any human interaction during testing, hence ZVOS is also known as unsupervised video object segmentation. Specifically, Zhou et al. [97] designed a Motion Attention Transmission (MAT) module in a dual-stream encoder network. By placing MAT in different convolutional layers, a tightly layered interaction between object appearance and motion is realized. They feed the modulated multi-scale spatio-temporal features into a boundary-aware decoder network to generate video object segmentation results. The COSNet [94] introduces the mechanism of co-attention in the visual question answering to unsupervised video object segmentation. The COSNet employs a Siamese network and uses paired video frames for training. Literature [95] improves COSNet by replacing the pairwise co-attention mechanism with a group co-attention mechanism.

Semi-supervised video object segmentation. In practical applications, users often want to segment specified or interested moving objects. At this time the user needs to provide certain target annotation information, the algorithm uses the annotation information provided by the user to segment the specified object in the subsequent video frames, which is semi-supervised video object segmentation. Earlier, researchers use interactive image segmentation algorithms to segment the objects of interest as annotation information in the first frame, and then use this annotation information to segment the specified objects of interest in the subsequent frames. However, the performance of different semi-supervised video object segmentation methods cannot be fairly evaluated because different researchers may use different interactive image segmentation algorithms to obtain the annotation mask for the first frame, and there are differences in the masks obtained by different interactive image segmentation techniques. Gradually, researchers use the ground-truth mask of the first frame as the guiding mask for the first frame, which becomes the standard paradigm for semi-supervised methods driven by the evaluation rules of the DAVIS dataset [98], also known as One-shot Video Object Segmentation (OVOS). In practice, it is time-consuming and laborious for users to manually annotate the target of the first frame of the video. Therefore, some researchers have started to study weakly supervised video object segmentation to improve the usability of semi-supervised video object segmentation methods. Specifically, researchers use a bounding box to specify the target to be segmented as supervised information in the first frame of the video, without the need to provide the accurate segmentation mask for the target in the first frame of the video. Furthermore, some researchers only provide semantic labels as supervisory information to specify the target to be segmented.

Interactive video object segmentation. The accuracy of segmentation results is increasingly required in applications such as video editing and action recognition, and the results of semi-supervised video object segmentation methods using only the first frame mask cannot meet the requirements. Therefore, interactive video object segmentation [99] is gradually developed, where users can continuously provide annotation information to improve the video object segmentation results. The annotation information can be given in the form of simple strokes or in the form of clicks, etc.

Multimodal video object segmentation. Multimodal video object segmentation is an important research direction in the cross field of natural language processing and computer vision. Referring video object segmentation [100] that combines two modalities, text and video, has gradually become an important task in video object segmentation. Specifically, a short text is used to give the information of the target to be segmented in the first frame, such as "a man riding on a horse", and then the algorithm segments the specified object according to the text information. Subsequently, the researchers further begin to study audio-visual segmentation [101], which requires algorithms to segment objects that emit sound in the video according to the audio files.

In summary, video object segmentation has attracted increasing attention from researchers and has undergone significant development. More challenging subtasks such as weakly supervised video object segmentation and referring video object segmentation have been proposed. The related algorithms can also serve different application requirements and are more convenient for practical applications. The related algorithms can also serve different application requirements and are more convenient for practical applications. In this book, we focus on the study of semi-supervised video object segmentation methods.

Existing video object segmentation methods do not fully utilize target structure information. The target structure information refers to the structural shape features of the target region itself, such as poles, ropes and other branch-like targets often contain straight-line regions, and targets such as cars and cargo boxes often contain right angles or corners. The target structure information serves as a crucial supplementary cue, augmenting appearance color features and spatio-temporal consistency information, particularly in intricate scenarios like target occlusion and articulated motion. This enhances the accuracy of video object segmentation algorithms. Nevertheless, current algorithms fail to fully harness the potential of structural features inherent in the targets. The convolutional network or Transformer module, while processing the entire image and extracting image features through a multi-layered approach, does not sufficiently account for the specific structural features of targets, including the shape structural information present in objects like rods, ropes, and cars. As the number of network layers deepens, the algorithm loses target structural features such as branches and corners. Therefore, how to fully utilize the target structure information is a key issue in achieving accurate video object segmentation.

To address the problem, we introduce graph convolutional networks into video object segmentation networks and propose a video object segmentation method based on structural and prototype feature extraction. We use the agglomerative clustering algorithm to obtain a series of sub-minimum spanning tree, which can make the spanning tree disconnected at the object boundaries and connected within the regions. The generated graph structure captures the target structural information, and the clustering operation in the feature space makes the graph generation process more robust. Further processing of graph data consisting of graph structure and video frame features using graph convolutional network enhances the target structural features. Meanwhile, in this book, we utilize the expectation maximum algorithm to solve the Gaussian mixture models, and adopt the resulting k-mean vectors as an aggregated representation of the target features to perform object-level matching, thus making full use of the target information. By introducing object-level matching, the prediction noise caused by point-to-point matching can be effectively reduced.

2.4 Challenges of Scene Understanding

Scene understanding is a series of computer vision and patter recognition tasks on image, video or 3D scene data. These tasks have faced different challenges from various factors. In order to build an efficient system to analyze these challenges, we divided them into 2 groups including internal factors and external factors according to their problem source, then every group has 4 sub-groups. The internal factors are from scene data themselves, such as changing of instance, structure complexity, object motion, data distribution and so on. The external factors are affected by external environment, human factors or physical device which captures the scene data, such as shooting perspective, camera configuration, shooting view environment variance and so on. In this section, we will introduce the external factors and internal interactions separately.

2.4.1 Internal and External Factors

Internal Factors include instance factors, structure factors, motion factors and data factors, as show in Fig. 2.5. The instance factors include appearance change, edge ambiguity, dense objects, interacting objects and heterogeneous objects. The structure factors include shape complexity and deformation. Motion factors include in-plane rotation, out-plane rotation, fast motion and motion blur. The data factors include category ambiguity, feature insufficiency and data imbalance.

The appearance change refers to noticeable variations in the appearance of an object caused by changes in illumination and the relative rotation between the camera and the object. The interacting object implies that the target object consists of multiple spatially-connected components, such as a man on horseback. Heterogeneous objects

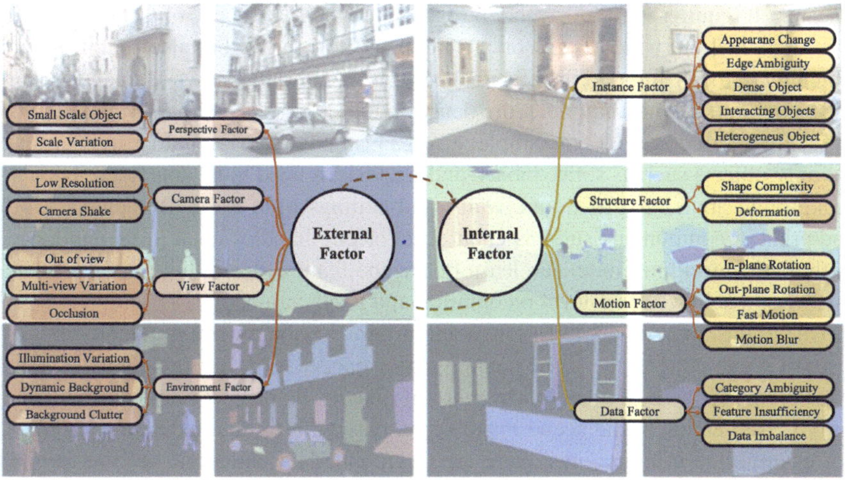

Fig. 2.5 Internal factors and external factors

mean that different regions of the object have distinct colors. The shape complexity indicates that the object has complex boundaries, including thin parts and holes. The deformation describes the object undergoing complex, non-rigid changes in shape. The motion blur refers to the object having fuzzy boundaries as a result of fast motion.

External Factors include perspective factors, camera factors, view factors and environment factors, as show in Fig. 2.5. The perspective factors include small scale object, scale variation. The camera factors include low resolution, camera shake. The view factors include out of view, multi-view variation and occlusion. The environment factors include illumination variation, dynamic background and background clutter.

The scale-variation is characterized by the area ratio between any pair of bounding boxes enclosing the target object being less than 0.5. The oow resolution is indicated by the ratio between the average object bounding box area and the image area being less than 0.1. The camera-shake results in noticeable vibrations in the footage. The out-of-view denotes that the object is partially clipped by the image boundaries. The occlusion means the object is partially or fully obscured. The dynamic background refers to background regions that move or deform. The background clutter describes the similarity in colors between the background and foreground regions around the object boundaries.

2.4.2 Challenges of Three Tasks

Scene Parsing is a more complex task compared to scene classification and scene attribute recognition, and it is also one of the focuses of future computer vision research. It requires algorithm models to achieve pixel level scene object recognition and segmentation, which is not only a multi label problem, but also requires label matching or localization to specific image pixels, commonly known as image semantic segmentation. This requires more dense annotation of scene images containing a large number of objects. For example, given a visual scene of an indoor living room, not only objects such as sofas, tables, and televisions need to be annotated, but also background areas such as floors, walls, and ceilings need to be segmented. More fine-grained components such as chair seats or cup handles need to be annotated, so that the algorithm model can learn fine scene features from the raw data. Scene parsing can achieve pixel-level recognition and localization.

Due to the complex composition, configuration attributes, and semantic relationships of scene images, the scene parsing performance is still insufficient to a certain extent. The richness of content, diversity of objects, diversity of layout, and ambiguity of semantics in the scene make scene parsing a more challenging problem. For example, in urban scenes, there are often abundant buildings, such as various tall buildings, low rise buildings, and special shaped buildings. Even simple scenes such as rivers and mountains often have specific spatial layouts. Figure 2.6 show that many water bodies are very similar, but they can serve as components of various different scenes, which leads to semantic ambiguity in some scene categories, such as oceans, rivers, estuaries. Therefore, scene recognition not only involves the existence of objects, but also the semantic relationship between objects and background information, resulting in significant inter class similarity and intra-class differences. In addition, there are certain difficulties in accurately representing complex semantic and spatial structural information. Although humans are very proficient in perceiving natural scenes and understanding advanced scene structures visually, solving this problem requires complex algorithm models and more visual features of images, such

Fig. 2.6 Examples of scene datasets

as shape, edges, texture, and color. ALl of this can effectively integrate information from multiple angles of a scene image, capture the interaction relationships between various scene objects, and obtain pixel level parsing results of the scene.

Human Parsing often referred to as semantic parts segmentation, is focused on the decomposition of the complete human body within an image into distinct body parts and clothing areas, each associated with independent semantics. These semantics typically include categories such as head, neck, coat, pants, and more. Similar to the task of semantic segmentation, human parsing involves a pixel-level classification task where the objective is to predict a semantic label for each pixel. Considering the internal factors, there are two types of hard samples for human parsing task, i.e., absence of limbs and complex body postures, as shown in Fig. 2.7. For absence of limbs, as shown in Fig. 2.7a, it refers to instances where parts of the human body, such as limbs (arms, legs), are not fully visible or are entirely absent in the image, especially in real-world scenarios where complete visibility of all body parts is not guaranteed. Addressing the absence of limbs is a significant challenge in human parsing because it requires the model to accurately predict the presence or absence of limbs based on the available information. For complex body postures, as shown in Fig. 2.7b, complex or unusual body postures make it challenging to accurately segment and parse the human body. Unconventional poses makes it difficult to create a universal model that accurately segments all possible variations. Some poses can be extreme which may involve body parts overlapping or being in atypical positions, and accurately parsing body parts in such poses is challenging.

Some external factors also cause serious errors in human parsing task, such as severe occlusions, dense crowd, scale and resolution variations. For severe occlusions, as shown in Fig. 2.7c, we mainly consider these cases which cause by the mutual occlusion between multiple bodies. In this case, occlusions occur when one body part is obstructed or partially hidden by another object or body part. Severe occlusions make it challenging to accurately delineate the boundaries of body parts and segment them correctly. For dense crowds, as shown in Fig. 2.7d, parsing human body in dense crowds can be challenging due to the overlapping of individuals and body parts. Distinguishing and segmenting each individual accurately in a crowded scene is a complex task, especially when body parts are partially obscured or merged with others. In addition, dense crowds poses a greater challenge to the instance-level human parsing. Scale and resolution variations is also a common external challenge for human parsing. Images may have variations in scale and resolution, making it necessary for the model to be robust to such differences. Adapting to different scales is crucial for accurately segmenting body parts. In addition, illumination and lighting conditions also causes serious interference to human parsing, which affects the model performance. Variations in lighting conditions, shadows, and highlights can impact the appearance of body parts. A robust human parsing model should be able to handle these lighting variations. For fine-grained human parsing, it needs to segment different styles of clothes, and different individuals may have variations in appearance and clothing styles. This need the human parsing model to be able to generalize across these differences. Hence, ambiguity and fine details also are challenging for human parsing.

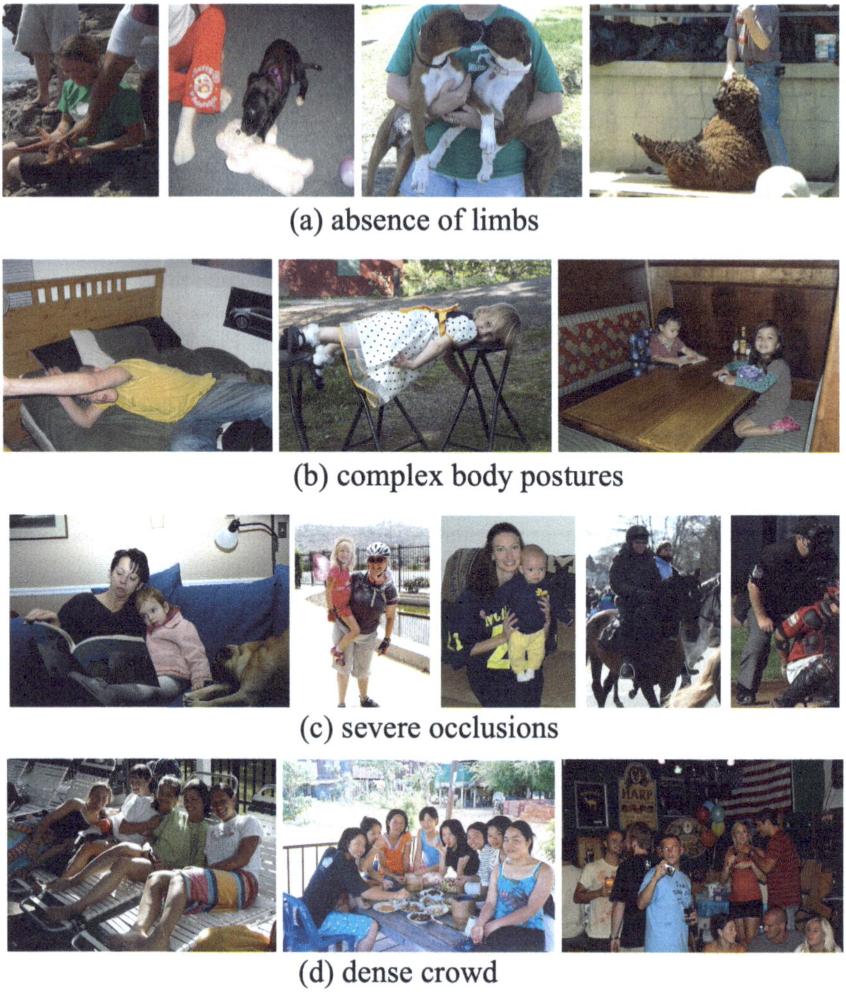

Fig. 2.7 Four types of hard samples for human parsing task

Addressing these challenges in human parsing is crucial for developing robust and accurate algorithms. It can handle a wide range of real-world scenarios, including those with varying poses, occlusions, and crowd densities. Researchers in computer vision continually work on improving algorithms to handle such complexities and enhance the performance of human parsing systems.

Video Object Segmentation needs to segment the object in the video data. The blue screen and green screen are both commonly used as backdrops for film and television shooting, where a video camera is used to capture clips of actors performing in front of a backdrop, which is then replaced with a specified background through post-production video editing techniques. This simple single background video object segmentation is relatively easy. However, there exist a wide variety of videos in practical applications, and the background is complex and variable. Perazzi et al. [98] classify the challenges posed by complex scenes into 15 categories including 7 external factors and 8 internal factors. Figure 2.8 illustrates five challenges including deformation and scale changes, the first column gives both the first frame of the video and the labeled segmentation mask.

The scale-variation, an external factor, means that the target object bounding box area ratio of any two frames is less than the threshold value $\tau_{sv} = 0.5$, such as the second line of Fig. 2.8 shows the Scale-Variation process of three targets from far to near. When the scale-variation occurs, it will lead to a large difference between the target objects in the subsequent frames of the video and the target objects in the first frame, and if the target objects in the first frame are smaller, the corresponding labeling information is less. The deformation, a internal factor, refers to the occurrence of complex non-rigid deformation of the target object, as shown in the video sequence in the first row of Fig. 2.8. The video sequence demonstrates the deformation that occurs during the dancer's dance.

In addition to the above challenges, there are also challenges such as background clutter, occlusion, low resolution, out-of-view, camera shake, shape complexity, motion blur, fast motion, appearance change, edge ambiguity, and heterogeneous objects. Among them, background clutter, occlusion, low resolution, out-of-view, and camera shake belong to external factors, while others belong to internal factors. The video sequence in the fifth row of Fig. 2.8 demonstrates the shape complexity challenge, which is a internal factor. There are many hollows inside the bicycle wheels

Fig. 2.8 Some examples of challenges, the first to fifth rows correspond in turn to deformation, scale-variation, background clutter, occlusion and shape complexity

in the video sequence, and the spokes in the wheels are dendritic regions. These complex dynamic scenes pose a great challenge to current video object segmentation algorithms.

This chapter mainly introduces the overview of scene parsing, taxonomy of scene understanding, three key tasks of this book, and the challenges of scene understanding. In the overview of scene parsing, from easy to hard, the tasks of scene understanding are divided into several levels, i.e., level-1, level-2, and level-3. Besides, according to scene data type, we also categorize scene understanding into image-based understanding, video-based understanding and 3D based understanding. Taxonomy of scene understanding are categorized into four classes, including scene representation, scene analysis, weak semantic scene understanding and strong semantic scene understanding. Following this, we detailedly introduce the three key tasks of this book, i.e., scene parsing, human parsing and video object segmentation. Finally, we summarize the challenges faced by the three main tasks in detail. Concretely, we classify these challenges into the external factors and internal interactions separately.

References

1. Xiao, Z., Deng, X., Lu, H., Xia, Y., Fan, X., Yi, L., Yang, L.T.: Tensor and confident information coverage based reliability evaluation for large-scale intelligent transportation wireless sensor networks. IEEE Transactions on Vehicular Technology 72(10), 13461–13473 (2023)
2. Huang, H., Shen, Z., Huang, C., Wang, Y., Wang, F.Y.: Intelligent vehicle carriers to support general civilian purposes. IEEE Transactions on Intelligent Vehicles 8(10), 4292–4295 (2023)
3. He, H., Gray, J., Cangelosi, A., Meng, Q., McGinnity, T.M., Mehnen, J.: The challenges and opportunities of human-centered ai for trustworthy robots and autonomous systems. IEEE Transactions on Cognitive and Developmental Systems 14(4), 1398–1412 (2022)
4. Ye, X., Wang, S., Gu, Y., Wang, J., Wang, R., Hou, B., Giunchiglia, F., Jiao, L.: A joint-training two-stage method for remote sensing image captioning. IEEE Transactions on Geoscience and Remote Sensing 60, 1–16 (2022)
5. Ni, P., Li, X., Kong, D., Yin, X.: Scene-adaptive 3d semantic segmentation based on multi-level boundary-semantic-enhancement for intelligent vehicles. IEEE Transactions on Intelligent Vehicles 9(1), 1722–1732 (2024)
6. Zhang, C., Ding, W., Peng, G., Fu, F., Wang, W.: Street view text recognition with deep learning for urban scene understanding in intelligent transportation systems. IEEE Transactions on Intelligent Transportation Systems 22(7), 4727–4743 (2021)
7. Tang, J., Liu, H.: Feature selection with linked data in social media. In: Proceedings of the International Conference on Data Mining, pp. 118–128 (2012)
8. Bengio, Y.: Neural net language models. Scholarpedia 3(1), 3881 (2008)
9. Krizhevsky, A., Sutskever, I., Hinton, G.E.: Imagenet classification with deep convolutional neural networks. In: Advances in Neural Information Processing Systems, pp. 1106–1114 (2012)
10. Simonyan, K., Zisserman, A.: Very deep convolutional networks for large-scale image recognition. arXiv preprint arXiv:1409.1556 (2014)
11. Szegedy, C., Liu, W., Jia, Y., Sermanet, P., Reed, S., Anguelov, D., Erhan, D., Vanhoucke, V., Rabinovich, A.: Going deeper with convolutions. In: Proceedings of the IEEE Conference on Computer Vision and Pattern Recognition, pp. 1–9 (2015)
12. He, K., Zhang, X., Ren, S., Sun, J.: Deep residual learning for image recognition. In: Proceedings of the IEEE Conference on Computer Vision and Pattern Recognition, pp. 770–778 (2016)

13. Huang, G., Liu, Z., van der Maaten, L., Weinberger, K.Q.: Densely connected convolutional networks. In: Proceedings of the IEEE Conference on Computer Vision and Pattern Recognition, pp. 2261–2269 (2017)
14. Gupta, P., Gupta, P.: Multi-modal fusion of palm-dorsa vein pattern for accurate personal authentication. Knowledge-Based Systems 81, 117–130 (2015)
15. Peng, W., Chen, A., Chen, J.: Using general master equation for feature fusion. Future Generation Computer Systems 82, 119–126 (2018)
16. Du, S., Wang, S.: An overview of correlation-filter-based object tracking. IEEE Transactions on Computational Social Systems 9(1), 18–31 (2022)
17. Zhang, L., Gonzalez-Garcia, A., van de Weijer, J., Danelljan, M., Khan, F.S.: Learning the model update for siamese trackers. In: Proceedings of the IEEE/CVF International Conference on Computer Vision, pp. 4009–4018 (2019)
18. Zhao, Z., Wang, J., Horn, M., Ding, Y., He, T., Bai, Z., Zietlow, D., Simon-Gabriel, C., Shuai, B., Tu, Z., Brox, T., Schiele, B., Fu, Y., Locatello, F., Zhang, Z., Xiao, T.: Object-centric multiple object tracking. In: Proceedings of the IEEE/CVF International Conference on Computer Vision, pp. 16601–16611 (2023)
19. Nguyen, H.V., Rezatofighi, H., Vo, B., Ranasinghe, D.C.: Distributed multi-object tracking under limited field of view sensors. IEEE Transactions on Signal Processing 69, 5329–5344 (2021)
20. Wang, Z., Xu, J., Liu, L., Zhu, F., Shao, L.: RANet: Ranking attention network for fast video object segmentation. In: Proceedings of the IEEE International Conference on Computer Vision, pp. 3977–3986 (2019)
21. Mao, Y., Wang, N., Zhou, W., Li, H.: Joint inductive and transductive learning for video object segmentation. In: Proceedings of the IEEE International Conference on Computer Vision, pp. 9650–9659 (2021)
22. Wang, Y., Zhao, Q., Wang, B., Wang, S., Zhang, Y., Guo, W., Feng, Z.: A real-time active pedestrian tracking system inspired by the human visual system. Cognitive Computation 8(1), 39–51 (2016)
23. Xu, X., Li, X., Zhao, H., Liu, M., Xu, A., Ma, Y.: A real-time, continuous pedestrian tracking and positioning method with multiple coordinated overhead-view cameras. Measurement 178, 109386 (2021)
24. Lu, S., Chen, T., Tian, S., Lim, J., Tan, C.L.: Scene text extraction based on edges and support vector regression. International Journal on Document Analysis and Recognition 18(2), 125–135 (2015)
25. Wang, C., Zhao, S., Zhu, L., Luo, K., Guo, Y., Wang, J., Liu, S.: Semi-supervised pixel-level scene text segmentation by mutually guided network. IEEE Transactions on Image Processing 30, 8212–8221 (2021)
26. Parseh, M.J., Rahmanimanesh, M., Keshavarzi, P., Azimifar, Z.: Semantic-aware visual scene representation. International Journal of Multimedia Information Retrieval 11(4), 619–638 (2022)
27. Cheng, G., Xie, X., Han, J., Guo, L., Xia, G.: Remote sensing image scene classification meets deep learning: Challenges, methods, benchmarks, and opportunities. IEEE Journal of Selected Topics in Applied Earth Observations and Remote Sensing 13, 3735–3756 (2020)
28. Danaci, E.G., Ikizler-Cinbis, N.: Low-level features for visual attribute recognition: An evaluation. Pattern Recognition Letters 84, 185–191 (2016)
29. Zeng, H., Song, X., Chen, G., Jiang, S.: Learning scene attribute for scene recognition. IEEE Transactions on Multimedia 22(6), 1519–1530 (2020)
30. Wang, S., Wang, Y., Zhu, S.: Learning hierarchical space tiling for scene modeling, parsing and attribute tagging. IEEE Transactions on Pattern Analysis and Machine Intelligence 37(12), 2478–2491 (2015)
31. Shi, Z., Yang, Y., Hospedales, T.M., Xiang, T.: Weakly-supervised image annotation and segmentation with objects and attributes. IEEE Transactions on Pattern Analysis and Machine Intelligence 39(12), 2525–2538 (2017)

32. Cheng, B., Li, Z., Xu, B., Dang, C., Deng, J.: Target detection in remote sensing image based on object-and-scene context constrained CNN. IEEE Geoscience and Remote Sensing Letters 19, 1–5 (2022)
33. Li, X., Ye, M., Liu, Y., Zhu, C.: Adaptive deep convolutional neural networks for scene-specific object detection. IEEE Transactions on Circuits and Systems for Video Technology 29(9), 2538–2551 (2017)
34. Du, L., Ling, H.: Dynamic scene classification using redundant spatial scenelets. IEEE Transactions on Cybernetics 46(9), 2156–2165 (2016)
35. Gu, Y., Liu, H., Wang, T., Li, S., Gao, G.: Deep feature extraction and motion representation for satellite video scene classification. Science China Information Sciences 63(4) (2020)
36. Zhang, Z., Zhang, L., Tong, X., Mathiopoulos, P.T., Guo, B., Huang, X., Wang, Z., Wang, Y.: A multilevel point-cluster-based discriminative feature for ALS point cloud classification. IEEE Transactions on Geoscience and Remote Sensing 54(6), 3309–3321 (2016)
37. Liao, L., Tang, S., Liao, J., Li, X., Wang, W., Li, Y., Guo, R.: A supervoxel-based random forest method for robust and effective airborne lidar point cloud classification. Remote Sensing 14(6), 1516 (2022)
38. Théodose, R., Denis, D., Chateau, T., Frémont, V., Checchin, P.: A deep learning approach for lidar resolution-agnostic object detection. IEEE Transactions on Intelligent Transportation Systems 23(9), 14582–14593 (2022)
39. Liu, H., Du, J., Zhang, Y., Zhang, H.: Extracting geometric and semantic point cloud features with gateway attention for accurate 3d object detection. Engineering Applications of Artificial Intelligence 123, 106227 (2023)
40. An, H., Hu, H.M., Guo, Y., Zhou, Q., Li, B.: Hierarchical reasoning network for pedestrian attribute recognition. IEEE Transactions on Multimedia 23, 268–280 (2021)
41. Fan, H., Hu, H.M., Liu, S., Lu, W., Pu, S.: Correlation graph convolutional network for pedestrian attribute recognition. IEEE Transactions on Multimedia 24, 49–60 (2022)
42. Xie, J., Pang, Y., Khan, M.H., Anwer, R.M., Khan, F.S., Shao, L.: Mask-guided attention network and occlusion-sensitive hard example mining for occluded pedestrian detection. IEEE Transactions on Image Processing 30, 3872–3884 (2021)
43. Lin, Z., Pei, W., Chen, F., Zhang, D., Lu, G.: Pedestrian detection by exemplar-guided contrastive learning. IEEE Transactions on Image Processing 32, 2003–2016 (2023)
44. Liu, J., Akhtar, N., Mian, A.: Adversarial attack on skeleton-based human action recognition. IEEE Transactions on Neural Networks and Learning Systems 33(4), 1609–1622 (2022)
45. Guo, F., Jin, T., Zhu, S., Xi, X., Wang, W., Meng, Q., Song, W., Zhu, J.: B2c-afm: bi-directional co-temporal and cross-spatial attention fusion model for human action recognition. IEEE Transactions on Image Processing 32, 4989–5003 (2023)
46. Wang, X., Liu, M., Raychaudhuri, D.S., Paul, S., Wang, Y., Roy-Chowdhury, A.K.: Learning person re-identification models from videos with weak supervision. IEEE Transactions on Image Processing 30, 3017–3028 (2021)
47. Yang, F., Zhong, Z., Luo, Z., Lian, S., Li, S.: Leveraging virtual and real person for unsupervised person re-identification. IEEE Transactions on Multimedia 22(9), 2444–2453 (2020)
48. Fu, Z., Chen, J., Jiang, K., Wang, S., Wen, J., Yang, M., Yang, D.: Traffic police 3d gesture recognition based on spatial–temporal fully adaptive graph convolutional network. IEEE Transactions on Intelligent Transportation Systems 24(9), 9518–9531 (2023)
49. Liu, X., Zhao, G.: 3d skeletal gesture recognition via discriminative coding on time-warping invariant riemannian trajectories. IEEE Transactions on Multimedia 23, 1841–1854 (2021)
50. Wei, P., Sun, H., Zheng, N.: Learning composite latent structures for 3d human action representation and recognition. IEEE Transactions on Multimedia 21(9), 2195–2208 (2019)
51. Li, M., Chen, S., Chen, X., Zhang, Y., Wang, Y., Tian, Q.: Symbiotic graph neural networks for 3d skeleton-based human action recognition and motion prediction. IEEE Transactions on Pattern Analysis and Machine Intelligence 44(6), 3316–3333 (2022)
52. Xu, Y., Wang, Y., Zhou, W., Wang, Y., Yang, Z., Bai, X.: Textfield: learning a deep direction field for irregular scene text detection. IEEE Transactions on Image Processing 28(11), 5566–5579 (2019)

53. Bi, H., Xu, C., Shi, C., Liu, G., Zhang, H., Li, Y., Dong, J.: Hgr-net: hierarchical graph reasoning network for arbitrary shape scene text detection. IEEE Transactions on Image Processing 32, 4142–4155 (2023)
54. Wu, L., Xu, Y., Hou, J., Chen, C.L.P., Liu, C.L.: A two-level rectification attention network for scene text recognition. IEEE Transactions on Multimedia 25, 2404–2414 (2023)
55. Li, M., Fu, B., Zhang, Z., Qiao, Y.: Character-aware sampling and rectification for scene text recognition. IEEE Transactions on Multimedia 25, 649–661 (2023)
56. Shi, H., Li, H., Meng, F., Wu, Q., Xu, L., Ngan, K.N.: Hierarchical parsing net: semantic scene parsing from global scene to objects. IEEE Transactions on Multimedia 20(10), 2670–2682 (2018)
57. Luo, A., Yang, F., Li, X., Li, Y., Jiao, Z., Cheng, H., Lyu, S.: Robust scene parsing by mining supportive knowledge from dataset. IEEE Trans. Neural Networks Learn. Syst. 34(5), 2633–2646 (2023)
58. Wang, X., Zhang, R., Shen, C., Kong, T., Li, L.: SOLO: A simple framework for instance segmentation. IEEE Transactions on Pattern Analysis and Machine Intelligence. 44(11), 8587–8601 (2022)
59. Yin, C., Tang, J., Yuan, T., Xu, Z., Wang, Y.: Bridging the gap between semantic segmentation and instance segmentation. IEEE Transactions on Multimedia 24, 4183–4196 (2022)
60. Cong, Y., Yang, M.Y., Rosenhahn, B.: Reltr: Relation transformer for scene graph generation. IEEE Transactions on Pattern Analysis and Machine Intelligence 45(9), 11169–11183 (2023)
61. Liu, D., Bober, M., Kittler, J.: Constrained structure learning for scene graph generation. IEEE Transactions on Pattern Analysis and Machine Intelligence 45(10), 11588–11599 (2023)
62. Miao, J., Wei, Y., Wu, Y., Liang, C., Li, G., Yang, Y.: Vspw: a large-scale dataset for video scene parsing in the wild. In: Proceedings of the IEEE Conference on Computer Vision and Pattern Recognition, pp. 4133–4143 (2021)
63. Kim, D., Woo, S., Lee, J., Kweon, I.S.: Dense pixel-level interpretation of dynamic scenes with video panoptic segmentation. IEEE Trans. Image Process. 31, 5383–5395 (2022)
64. Rother, C., Kolmogorov, V., Blake, A.: "grabcut": interactive foreground extraction using iterated graph cuts. ACM Trans. Graph. 23(3), 309–314 (2004)
65. Zhang, S., Liew, J.H., Wei, Y., Wei, S., Zhao, Y.: Interactive object segmentation with inside-outside guidance. In: Proceedings of the IEEE Conference on Computer Vision and Pattern Recognition, pp. 12231–12241 (2020)
66. Xu, C., Dong, B., Stier, N., McCully, C., Howell, D.A., Sen, P., Höllerer, T.: Interactive segmentation and visualization for tiny objects in multi-megapixel images. In: Proceedings of the IEEE Conference on Computer Vision and Pattern Recognition, pp. 21415–21420 (2022)
67. Nie, D., Lan, R., Wang, L., Ren, X.: Pyramid architecture for multi-scale processing in point cloud segmentation. In: Proceedings of the IEEE Conference on Computer Vision and Pattern Recognition, pp. 17263–17273 (2022)
68. Shuai, H., Liu, Q.: Geometry-injected image-based point cloud semantic segmentation. IEEE Transactions on Geoscience and Remote Sensing 61, 1–10 (2023)
69. Liu, X., Zhao, Y., Zhu, S.C.: Single-view 3d scene reconstruction and parsing by attribute grammar. IEEE Transactions on Pattern Analysis and Machine Intelligence 40(3), 710–725 (2018)
70. Zhang, S., Cao, X., Qi, G.J., Song, Z., Zhou, J.: Aiparsing: anchor-free instance-level human parsing. IEEE Transactions on Image Processing 31, 5599–5612 (2022)
71. Ke, L., Chang, M.C., Qi, H., Lyu, S.: Detposenet: improving multi-person pose estimation via coarse-pose filtering. IEEE Transactions on Image Processing 31, 2782–2795 (2022)
72. Fang, H.S., Li, J., Tang, H., Xu, C., Zhu, H., Xiu, Y., Li, Y.L., Lu, C.: Alphapose: whole-body regional multi-person pose estimation and tracking in real-time. IEEE Transactions on Pattern Analysis and Machine Intelligence 45(6), 7157–7173 (2023)
73. Lee, K., Kim, W., Lee, S.: From human pose similarity metric to 3d human pose estimator: temporal propagating lstm networks. IEEE Transactions on Pattern Analysis and Machine Intelligence 45(2), 1781–1797 (2023)

74. Zhang, J., Gong, K., Wang, X., Feng, J.: Learning to augment poses for 3d human pose estimation in images and videos. IEEE Transactions on Pattern Analysis and Machine Intelligence 45(8), 10012–10026 (2023)
75. Farenzena, M., Bazzani, L., Perina, A., Murino, V., Cristani, M.: Person re-identification by symmetry-driven accumulation of local features. In: Proceedings of the IEEE Conference on Computer Vision and Pattern Recognition, pp. 2360–2367 (2010)
76. Wang, Y., Tran, D., Liao, Z., Forsyth, D.A.: Discriminative hierarchical part-based models for human parsing and action recognition. Journal of Machine Learning Research 13, 3075–3102 (2012)
77. Yamaguchi, K., Kiapour, M.H., Berg, T.L.: Paper doll parsing: Retrieving similar styles to parse clothing items. In: Proceedings of the IEEE International Conference on Computer Vision, pp. 3519–3526 (2013)
78. Wang, W., Xu, Y., Shen, J., Zhu, S.: Attentive fashion grammar network for fashion landmark detection and clothing category classification. In: Proceedings of the IEEE Conference on Computer Vision and Pattern Recognition, pp. 4271–4280 (2018)
79. He, K., Zhang, X., Ren, S., Sun, J.: Deep residual learning for image recognition. In: Proceedings of the IEEE Conference on Computer Vision and Pattern Recognition, pp. 770–778. IEEE Computer Society (2016)
80. Ren, S., He, K., Girshick, R.B., Sun, J.: Faster R-CNN: towards real-time object detection with region proposal networks. In: Neural Information Processing Systems, pp. 91–99 (2015)
81. Everingham, M., Gool, L.V., Williams, C.K.I., Winn, J.M., Zisserman, A.: The pascal visual object classes (VOC) challenge. International Journal of Computer Vision 88(2), 303–338 (2010)
82. Russakovsky, O., Deng, J., Su, H., Krause, J., Satheesh, S., Ma, S., Huang, Z., Karpathy, A., Khosla, A., Bernstein, M.S., Berg, A.C., Fei-Fei, L.: Imagenet large scale visual recognition challenge. International Journal of Computer Vision 115(3), 211–252 (2015)
83. Taigman, Y., Yang, M., Ranzato, M., Wolf, L.: Deepface: Closing the gap to human-level performance in face verification. In: Proceedings of the IEEE Conference on Computer Vision and Pattern Recognition, pp. 1701–1708. IEEE Computer Society (2014)
84. Cao, Z., Simon, T., Wei, S., Sheikh, Y.: Realtime multi-person 2d pose estimation using part affinity fields. In: Proceedings of the IEEE Conference on Computer Vision and Pattern Recognition, pp. 1302–1310. IEEE Computer Society (2017)
85. Zhao, Y., Li, J., Zhang, Y., Tian, Y.: From pose to part: Weakly-supervised pose evolution for human part segmentation. IEEE Transactions on pattern analysis and machine intelligence 45(3), 3107–3120 (2023)
86. Fang, H., Lu, G., Fang, X., Xie, J., Tai, Y., Lu, C.: Weakly and semi supervised human body part parsing via pose-guided knowledge transfer. In: Proceedings of the IEEE Conference on Computer Vision and Pattern Recognition, pp. 70–78 (2018)
87. Liu, Y., Zhang, S., Yang, J., Yuen, P.C.: Hierarchical information passing based noise-tolerant hybrid learning for semi-supervised human parsing. In: Proceedings of AAAI Conference on Artificial Intelligence, pp. 2207–2215 (2021)
88. Zhou, T., Porikli, F., Crandall, D.J., Van Gool, L., Wang, W.: A survey on deep learning technique for video segmentation. IEEE Transactions on Pattern Analysis and Machine Intelligence 45(6), 7099–7122 (2023)
89. Yang, Z., Wei, Y., Yang, Y.: Collaborative video object segmentation by multi-scale foreground-background integration. IEEE Transactions on Pattern Analysis and Machine Intelligence 44(9), 4701–4712 (2022)
90. Wang, M., Mei, J., Liu, L., Tian, G., Liu, Y., Pan, Z.: Delving deeper into mask utilization in video object segmentation. IEEE Trans. Image Process. 31, 6255–6266 (2022)
91. Guo, P., Zhang, W., Li, X., Zhang, W.: Adaptive online mutual learning bi-decoders for video object segmentation. IEEE Transactions on Image Processing 31, 7063–7077 (2022)
92. Wang, W., Lu, X., Shen, J., Crandall, J.D., Shao, L.: Zero-shot video object segmentation via attentive graph neural networks. In: Proceedings of the IEEE International Conference on Computer Vision, pp. 9236–9245 (2019)

93. Lu, X., Wang, W., Shen, J., Crandall, D.J., Van Gool, L.: Segmenting objects from relational visual data. IEEE Transactions on Pattern Analysis and Machine Intelligence 44(11), 7885–7897 (2022)
94. Lu, X., Wang, W., Ma, C., Shen, J., Shao, L., Porikli, F.: See more, know more: Unsupervised video object segmentation with co-attention siamese networks. In: Proceedings of the IEEE Conference on Computer Vision and Pattern Recognition, pp. 3618–3627 (2019)
95. Lu, X., Wang, W., Shen, J., Crandall, D., Van Gool, L.: Zero-shot video object segmentation with co-attention siamese networks. IEEE Transactions on Pattern Analysis and Machine Intelligence 44(4), 2228–2242 (2022)
96. Lu, X., Wang, W., Shen, J., Tai, Y.W., Crandall, D.J., Hoi, S.C.H.: Learning video object segmentation from unlabeled videos. In: Proceedings of the IEEE Conference on Computer Vision and Pattern Recognition, pp. 8957–8967 (2020)
97. Zhou, T., Li, J., Wang, S., Tao, R., Shen, J.: MATNet: Motion-attentive transition network for zero-shot video object segmentation. IEEE Transactions on Image Processing 29, 8326–8338 (2020)
98. Perazzi, F., Pont-Tuset, J., Mcwilliams, B., Gool, L.V., Gross, M., Sorkine-Hornung, A.: A benchmark dataset and evaluation methodology for video object segmentation. In: Proceedings of the IEEE Conference on Computer Vision and Pattern Recognition, pp. 724–732 (2016)
99. Cheng, H.K., Tai, Y.W., Tang, C.K.: Modular interactive video object segmentation: Interaction-to-mask, propagation and difference-aware fusion. In: Proceedings of the IEEE Conference on Computer Vision and Pattern Recognition, pp. 5555–5564 (2021)
100. Khoreva, A., Rohrbach, A., Schiele, B.: Video object segmentation with language referring expressions. In: Proceedings of the Asian Conference on Computer Vision (2018)
101. Zhou, J., Wang, J., Zhang, J., Sun, W., Zhang, J., Birchfield, S., Guo, D., Kong, L., Wang, M., Zhong, Y.: Audio-visual segmentation. In: Proceedings of the European Conference on Computer Vision (2022)

Chapter 3
Graph Neural Network Basis

The graph structure is generally composed of nodes and edges, and the nodes and edges can carry different information respectively. Since classical machine learning algorithms usually can only take the original input data of vectors, a kind of algorithm model that can directly take the graph data as input—graph neural network is proposed.

3.1 The General GNN Framework

We first introduce the general graph neural network framework using node-level tasks as an example. Then we introduce the general framework for graph-level tasks, which require the additional use of graph pooling operations in comparison to node-level tasks.

Taking the node-level task as an example, a generic framework mainly contains feature transformation layers and nonlinear activation layers. The network usually contains multiple feature transformation layers, and the i-th feature transformation layer is shown in the following equation:

$$H^{(i)} = \mathrm{Ftr}_i\left(A, \sigma_{i-1}\left(H^{(i-1)}\right)\right), \tag{3.1}$$

where $H^{(i-1)}$ remarks the hidden layer features after the $(i-1)$th feature transformation layer. σ_{i-1} remarks the nonlinear activation layer. A is the adjacency matrix. The i-th feature transformation layer is indicated by Ftr_i. $H^{(i)} \in \mathbb{R}^{N \times d_i}$, N denotes the number of graph nodes and d_i denotes the dimension of the i-th feature transformation layer output. Some special layers are employed to process the final $H^{(L)}$ to accomplish different node-level tasks.

W. Liu et al., *Graph Neural Network Methods and Applications in Scene Understanding*,
https://doi.org/10.1007/978-981-97-9933-6_3

As for node classification, we can use softmax to get the predicted output $\hat{y}_v = \text{softmax}\left(PH_v^T\right)$ of node v. $P \in \mathbb{R}^{C\times d}$ and C denotes the number of categories of the label. d indicates the dimension of the node features. The loss function is $\mathcal{L} = \frac{1}{N}\sum_{k=1}^{N} \text{loss}\left(\hat{y}_k, y_k\right)$, where y_k remarks the ground-truth label corresponding to node k. $\text{loss}(\cdot)$ denotes the loss function, usually using the cross-entropy (CE) loss function.

For graph-level tasks, a general framework requires graph pooling layers in addition to feature transformation layers and nonlinear activation layers. The feature transformation layers and the nonlinear activation layers are also utilized to extract better node features. The graph pooling layer allows us to obtain overall information about the graph. The graph pooling layer summarizes information from multiple nodes and produces a coarser representation of the graph, which can also help in reducing the computational complexity of subsequent layers and retain important structural information. For a network block containing a graph pooling layer, the graph pooling layer is usually located in the last layer of the block, and the other layers in the block are a series of feature transformation layers and nonlinear activation layers. We use $H^{(bi)}$ and $A^{(bi)}$ to denote the features and adjacency matrix of the graph data $G_{in} = (V_{bi}, E_{bi})$ input to the block, respectively. The coarsened graph is obtained from the following equation:

$$H^{(i)} = \text{Ftr}_i\left(A^{(bi)}, \sigma_{i-1}\left(H^{(i-1)}\right)\right) \, for \quad i = 1, 2, \ldots, m \tag{3.2}$$

$$A^{(bo)}, H^{(bo)} = \text{Pool}\left(A^{(bi)}, H^{(i-1)}\right), \tag{3.3}$$

where $H^{(bo)}$ and $A^{(bo)}$ denote the features and adjacency matrix of the graph data $G_{bo} = (V_{bo}, E_{bo})$ output by the module, respectively. $H^{(0)} = H^{(bi)}$ and Pool remarks the pooling layer. The GNN framework can stack multiple such blocks.

Further, the feature transformation layer can be divided into two steps of aggregation and combination as shown in the following equation:

$$a_v^{(i)} = \text{AGGREGATE}^i\left(\left\{H_u^{(i)}, \forall u \in \mathcal{N}(v)\right\}\right), \tag{3.4}$$

$$H_v^{\prime(i+1)} = \text{COMBINE}^i\left(H_v^{(i)}, a_v^{(i)}\right), \tag{3.5}$$

where $\mathcal{N}(v)$ denotes the neighborhood of node v. AGGREGATE^i is performed to obtain the information $a_v^{(i)}$ aggregated from neighboring nodes $\mathcal{N}(v)$. i denotes the index of the feature transformation layer. COMBINE^i is performed to obtain the updated node representation $H_v^{(i+1)}$. $H_v^{\prime(i+1)}$ undergoes a nonlinear activation layer to obtain the $H_v^{(i+1)}$.

Above we have introduced the generalized framework of graph neural networks. Typical operations of graph neural networks are described in the following sections.

3.2 Operations of Graph Neural Networks

Gori et al. [1] pioneeringly employ the graph neural network model to introduce this novel methodology, thereby coining the term for this class of approaches. Gori et al. propose the idea of using the state of neighbor nodes for data learning in the graph domain. The attribute information and topological relationship carried by graph data nodes (or edges) are encoded as low-dimensional vectors after being processed by GNN model, and literature [2] improves the GNN theoretical model. The recursive neural network is used as the internal encoding network to learn the node state, and the differentiability of the model and the back propagation algorithm training are discussed, and the application problems of the model are further expounded. With the research, many graph neural networks such as spectral graph convolutional neural network and spatial graph convolutional network have been developed. In these networks, the main operations are spectral graph convolution, spatial graph convolution, graph attention, etc.

3.2.1 Spectral Graph Convolution

Spectral graph convolution is an operation of graph signal processing defined on graph Laplacian matrix in the frequency domain. The Laplacian matrix of the graph can be expressed as $\mathbf{L} = \mathbf{D} - \mathbf{A}$, where $\mathbf{D}$ is the diagonal matrix of node degree and $\mathbf{A}$ is the adjacency matrix of the graph. The Laplacian spectrum is the eigenvalue of $\mathbf{L}$. The normalized Laplacian matrix is defined as $\mathbf{L} = I_n - \mathbf{D}^{-\frac{1}{2}}\mathbf{A}\mathbf{D}^{-\frac{1}{2}}$, where $\mathbf{D}_{ii} = \sum_j \mathbf{A}_{i,j}$, $\mathbf{L}$ is a real semi-positive definite symmetric matrix, which can be obtained by eigenvalue decomposition (EVD) as follows.

$$\mathbf{L} = U\bar{\mathbf{\Lambda}}U^{\mathrm{T}}, \tag{3.6}$$

where $\bar{\mathbf{\Lambda}}$ is the eigenvalue diagonal matrix (i.e. spectrum), and U is the eigenvector matrix.

Based on the above graph signal processing method, the feature fusion is realized by spectral graph convolution operation. Firstly, let $\boldsymbol{x} \in \mathbb{R}^{\mathbb{C}}$, and define the Fourier transform (Eq. (3.7)) and inverse Fourier transform (Eq. (3.8)) of the graph as follows,

$$F(\boldsymbol{x}) = U^{\mathrm{T}}\boldsymbol{x}, \tag{3.7}$$

$$F^{-1}(\hat{\boldsymbol{x}}) = U^{\mathrm{T}}\hat{\boldsymbol{x}}, \tag{3.8}$$

where U is the eigenvector matrix obtained by EVD in the Eq. (3.6). The input raw graph data is projected into an orthogonal space with base U^{T} by Fourier transform, denoted as $\boldsymbol{x} = \sum_i \hat{x}_i u_i$. Therefore, the spectral graph convolution operation for input signal $\boldsymbol{x}$ by filter g can be defined as

$$x *_G g = F^{-1}(F(x) \odot F(g)) = U(U^{\mathrm{T}}\boldsymbol{x} \odot U^{\mathrm{T}}g), \tag{3.9}$$

where $*_G$ denotes spectral graph convolution, $\odot$ is element-wise multiplication. Then let $g_\theta = diag(U^{\mathrm{T}}g)$, Eq. (3.9) can be simplified to

$$x *_G g_\theta = Ug_\theta U^{\mathrm{T}}\boldsymbol{x}, \tag{3.10}$$

where, let $g_\theta = \Theta_{i,j}^{(t)}$ and $\Theta_{i,j}^{(t)}$ is a diagonal matrix. $\Theta_{i,j}^{(t)}$ is a learnable parameters set. Then, the (t)th layer of spectral graph convolution can be defined as

$$H_{:,j}^{(k)} = \sigma(\sum_{i=1}^{\mathbb{C}_{(t-1)}} U\Theta_{i,j}^{(t)}U^{\mathrm{T}}H_{:,i}^{(t-1)}),\ (j = 1, 2, \ldots, \mathbb{C}_{(t)}), \tag{3.11}$$

where $H^{(0)} = X$, $\mathbb{C}_{(t)}$ and $\mathbb{C}_{(t-1)}$ are the number of output channels of (t)th and $(t-1)$th layer, $\sigma(\cdot)$ is the activation function. The eigenvalue matrix $\bar{\mathbf{\Lambda}}$ and eigenvector matrix U in the Eq. (3.6) are sensitive to the change of graph structure, which can be learned by g_θ from the Eq. (3.10). Using this method, the edge weights are updated in backward calculation.

3.2.2 Spatial Graph Convolution

In the graph convolutional network model, spatial graph convolution module has the effect of feature learning. The neighboring nodes of a node can be classified into K levels. The nearest neighbor is level-1, the neighbor of the nearest neighbor is level-2, and so on. Through collecting message from multi-level neighbors, every node can obtain more information for updating model weight. According to different edge types, one node may have spatial neighbor connected by spatial edge or spatial hierarchical neighbor connected by spatial hierarchical edge. Both or one of the neighbors can be used for convolutional operation. An output state of the node v at model layer (t) is shown in the Eq. (3.12) without considering the activation function temporarily.

$$\mathbf{h}_v^{(t)} = (\boldsymbol{w}_v^{\mathrm{T}})^{(t)}\boldsymbol{x}_v + \sum_{k=1}^{\mathrm{K}} \sum_{u \in \mathcal{N}_k(v)} (\boldsymbol{\theta}_k^{\mathrm{T}})^{(t)}\mathbf{h}_u^{(t-1)}, \tag{3.12}$$

where $\mathbf{h}_v^{(0)} = 0$, and $\boldsymbol{w}_v$ indicates the attribute weight of node v, $\boldsymbol{\theta}$ indicates the edge connection weight of node v; K is a constant, which denotes the K-level neighbor of node v, then $\mathcal{N}_k(v)$ is set of level-k neighbor of node v. Its matrix form can be written as the Eq. (3.13).

$$H^{(t)} = XW^{(t)} + \sum_{k=1}^{K} \mathbf{A}H^{(t-1)}\Theta_k^{(t)}, \tag{3.13}$$

where $\mathbf{A}$ is the adjacency matrix, and Θ_k is the edge connection weight matrix of the node neighbor of the layer.

As shown in Eq. (3.14), at this time, attributes of the original data are learned directly through spatial graph convolution operation.

$$\mathbf{h}_v^{(0)} = (\boldsymbol{w}_v^{\mathrm{T}})^{(0)}\boldsymbol{x}_v. \tag{3.14}$$

Based on message passing mechanism, convolution operation process of Eq. (3.12) and node state update process can be reformulated as message passing and state update process. The message function Message(·) and the update function Update(·) are expressed by the Eqs. (3.15) and (3.16).

$$\mathbf{m}_v = \sum_{k=1}^{K} \sum_{u \in \mathcal{N}_k(v)} \mathrm{Message}(\mathbf{h}_u). \tag{3.15}$$

$$\mathbf{x}_v' = \mathrm{Update}(\mathbf{x}_v, \mathbf{m}_v). \tag{3.16}$$

Using Eq. (3.12) can deduce a new node attribute $\boldsymbol{x}_v' \in \mathbb{R}^{\mathbb{C}'}$. For fixed graph structure, complete state output can be represented as $X' \in \mathbb{R}^{\mathbb{N}\times\mathbb{L}\times\mathbb{C}'}$. Therefore, based on the convolutional operation, there is a feature mapping $F : X \rightarrow X'$, which implements high-level feature learning of each node. And the feature will be node representation for downstream tasks.

3.2.3 Graph Attention

Due to the global sharing of the graph structure in the network model, the message passing process of the Eqs. (3.15) and (3.16) can only proceed along the predefined path of $\mathbf{A}$. It means that the model will carry out the same efficiency of feature learning for nodes with key attributes and nodes with noise, which is not conducive to model training and tuning.

In order to improve the efficiency of feature learning, a graph attention module can be designed by parameterized self-supervised attention mechanism, introducing certain dynamic changes to the network model to make the nodes within K-level neighbor no longer pass message with the same weight.

A weighted adjacency matrix is used to implement the self-supervised attention mechanism, and the weight of connection represents the intensity of attention. Given node v, assuming that a node u contains key attributes in its neighbor, then it should strengthen the attention of v to u, which means that there is a stronger connection between v and u than other nodes. Let $e_{u,v} \in (0, 1)$ denote the relationship, a larger $e_{u,v}$ indicates stronger attention. Given node features $\boldsymbol{x}_u$ and $\boldsymbol{x}_v$, let $\Phi \in \mathbb{R}^{\mathbb{C}\times\mathbb{C}}$ be a learning parameter matrix. Then the graph attention mechanism could be defined as a mapping $a : \mathbb{R}^{\mathbb{C}} \times \mathbb{R}^{\mathbb{C}} \to \mathbb{R}$ shown in the Eq. (3.17),

$$\begin{aligned} e_{u,v} &= \mathrm{softmax}_{u\in\mathcal{N}(v)}(\Phi\boldsymbol{x}_u, \Phi\boldsymbol{x}_v) \\ &= \frac{\exp(||\Phi\boldsymbol{x}_v||)}{\sum_{u\in\mathcal{N}(v)}\exp(||\Phi\boldsymbol{x}_u||)}, \end{aligned} \tag{3.17}$$

where $||\cdot||$ is the vector norm. Further, the Eqs. (3.12) and (3.13) can be adjusted to the following Eqs. (3.18) and (3.19), respectively.

$$\mathbf{h}_v^{(t)} = (\boldsymbol{w}_v^{\mathrm{T}})^{(t)}\boldsymbol{x}_v + \sum_{k=1}^{\mathrm{K}}\sum_{u\in\mathcal{N}_k(v)}(\boldsymbol{\theta}_k^{\mathrm{T}})^{(t)}\mathbf{h}_u^{(t-1)}e_{u,v}. \tag{3.18}$$

$$H^{(t)} = XW^{(t)} + \sum_{k=1}^{\mathrm{K}}\mathbf{A}_\Phi^{(t-1)}H^{(t-1)}\Theta_k^{(t)}. \tag{3.19}$$

This mechanism can achieve different attention intensity by strengthening or weakening local connections. It does not change the number of nodes or edges, but changes the weight of connections between nodes, which means that the value in the adjacency matrix $\mathbf{A}_\Phi$ is no longer discrete, but is $e_{u,v} \in (0, 1)$ in the continuous domain. It makes the node with key attributes play a more important role in information transmission, so that the network model can learn the node more effectively. In addition, this mechanism also plays a certain inhibitory role on network connections, avoiding the simultaneous activation of too many connections in the network. On one hand, it can accelerate model training, and on the other hand, it can alleviate the problem of over-fitting.

3.2.4 Graph Pooling

A graph pooling module can reduce redundant information and speed up network model training. Pooling the graph structure will change the internal adjacency matrix constructed by the graph construction on scene images, which is mainly manifested by the reduction of nodes or edges, and it will obtain a new graph structure. The graph structure pooling is formulated as:

$$G^{'} = \mathrm{pool}_G(G), \tag{3.20}$$

where, G is the original graph structure. Maximum pooling $\text{pool}_{\max}(\cdot)$ and average pooling $\text{pool}_{\text{avg}}(\cdot)$ are two of the most commonly used pooling methods to get information about nodes that represent the whole graph or subgraph.

In recent years, researchers have proposed a variety of graph pooling methods [3–5] to accommodate different application scenarios and dataset characteristics. Gao et al. proposed the Top-k node selection method [3]. The authors use a trainable projection vector to project all node features into 1 dimension and then selected the largest k nodes according to the pooling ratio to obtain a new adjacency matrix and feature matrix.

Bianchi et al. [4] propose a clustering function that can be quickly evaluated on out-of-sample graphs. Based on the proposed clustering method, the authors design a graph pooling operator minCUT. The operator is based on GNN implementation and differentiable and does not require the computation of spectral decomposition. More graph pooling methods can be found in the review article on graph pooling by Grattarola et al. [6].

3.3 Architectures of Graph Neural Networks

Graph Neural networks (GNNs) are a class of methods that use neural networks to learn graph-structured data. GNNs can learn vector representations of nodes based on graph structure and node characteristics. Suppose the graph structure is, the node eigen matrix is, and we want to learn the node vector matrix. The basic idea of GNN is to define the neural network directly using the graph structure, and to define the updating mode of the node vector representation based on the neural network. The parameters of neural networks are usually global, that is, shared for every node in the graph, so GNN is more computationally efficient than shallow vector methods. In this section, we introduce the representative work of three GNNs: original Graph Neural Networks, Graph Convolutional Networks (GCNs), and Graph Attention Networks (GATs).

3.3.1 Graph Neural Network

Graph Neural Networks (GNNs) are a deep learning method based on graph structure, which have been widely used in various tasks such as image and natural language processing. As an extension of neural networks, graph neural networks can handle data formats represented by graph structures. In the graph, each node is defined by its own characteristics as well as its neighbors and relations, and the network computes the representation vector of the node by recursively aggregating and transforming the representation vector of the neighboring nodes.

In Relational inductive biases, deep learning, and graph networks, the author defines the graph networks (GN) framework. It generalizes and extends various graph neural networks, and supports the use of simple modules to build complex structures. The main computing unit of the GN framework is the GN block, a graph-to-graph module, whose input is a graph, which is calculated on the graph structure, and then gets the output of the same graph structure.

GNNs have flexible structure and update methods, which can well express the structural characteristics of some data itself. In addition to some data sets with graph structure (such as Cora, Citeseer, etc.), graph neural networks have also been applied to more tasks, such as text summarization, text classification and sequence annotation. So far, GNNs and their variants have achieved the best results for many tasks.

We introduce the development of GNNs in Chap. 1. In this section, we introduce a typical graph neural network method Gated GNN [7]. Gated GNN employs Gated Recurrent Units (GRU). The number of iterative update steps is fixed. Backpropagation by time is utilized in computing the gradient.

In the original GNN [2], the state of the node after convergence is independent of the initialization. The Gated GNN is dependent on the initial input of the node. For a reachability task, we need to predict whether we can reach node t from node s. The Gated GNN can receive annotation information. The annotation information for node t is $\boldsymbol{x}_t = [0, 1]^\top$. The annotation information for node s is $\boldsymbol{x}_s = [1, 0]^\top$. $\boldsymbol{x}_u = [0, 0]^\top$ indicates annotation information for other nodes u. These annotations are used to initialize the node states $\mathbf{h}_u^{(1)}$, while zeros can be padded in for higher dimensions. The model only needs to determine whether the first two dimensions of the node t vector have non-zero values in the final stage to predict reachability. The Gated GNN divides the network into propagation and output layers.

The output layer employs operations in a recurrent neural network, as shown in the following equation:

$$\mathbf{h}_u^{(1)} = \left[\boldsymbol{x}_u^\top, \mathbf{0}\right]^\top, \tag{3.21}$$

$$\mathbf{a}_u^{(t)} = \mathbf{A}_{u:}^\top \left[\mathbf{h}_1^{(t-1)\top} \ldots \mathbf{h}_{|V|}^{(t-1)\top}\right]^\top + \mathbf{b}, \tag{3.22}$$

$$\mathbf{z}_u^t = \sigma\left(\mathbf{W}^z \mathbf{a}_u^{(t)} + \mathbf{U}^z \mathbf{h}_u^{(t-1)}\right), \tag{3.23}$$

$$\mathbf{r}_u^t = \sigma\left(\mathbf{W}^r \mathbf{a}_u^{(t)} + \mathbf{U}^r \mathbf{h}_u^{(t-1)}\right), \tag{3.24}$$

$$\widetilde{\mathbf{h}_u^{(t)}} = \tanh\left(\mathbf{W}\mathbf{a}_u^{(t)} + \mathbf{U}\left(\mathbf{r}_u^t \odot \mathbf{h}_u^{(t-1)}\right)\right), \tag{3.25}$$

$$\mathbf{h}_u^{(t)} = \left(1 - \mathbf{z}_v^t\right) \odot \mathbf{h}_u^{(t-1)} + \mathbf{z}_u^t \odot \widetilde{\mathbf{h}_u^{(t)}}, \tag{3.26}$$

where, $\mathbf{A} = \left[\mathbf{A}^{(\text{out})}, \mathbf{A}^{(\text{in})}\right]$ and $\mathbf{A} \in \mathbb{R}^{D|V| \times 2D|V|}$. $\mathbf{A}$ represents the connection between nodes and is sparse. The elements in $\mathbf{A}$ indicate the type and direction of the edge. Equation (3.21) represents the initialization operation. Equation (3.22) represents the information transfer between nodes. $\mathbf{a}_u^{(t)} \in \mathbb{R}^{2D}$ is the aggregate information obtained. $\mathbf{z}$ and $\mathbf{r}$ represent update and reset gates, respectively, calculated by Eqs. (3.23) and (3.24). $\odot$ denotes the element multiplication. σ is the sigmoid activation function. It can be seen that the hidden layer feature $\mathbf{h}$ is updated in a GRU-like manner.

In the output layer, for the node selection task, one can make $o_u = g(\mathbf{h}_u^{(T)}, \boldsymbol{x}_u)$, i.e., get the node score through the fully connected layer and the softmax operation to select the node. For graph-level tasks, one can use Eq. (3.27) to obtain graph-level representation vectors, and then process them at the output layer.

$$\mathbf{h}_{\mathcal{G}} = \tanh\left(\sum_{u \in V} \sigma\left(s_1(\mathbf{h}_u^{(T)}, \boldsymbol{x}_u)\right) \odot \tanh\left(s_2(\mathbf{h}_u^{(T)}, \boldsymbol{x}_u)\right)\right), \tag{3.27}$$

where $\sigma(\cdot)$ is used to output the importance weights of nodes. s_1 and s_2 are both neural networks, and the inputs of the two networks are concatenated first. The outputs of both networks are real valued vectors.

The above model is referred to here as the initial model, and the authors obtain a serialized model based on the initial model to produce the serialized output $\boldsymbol{O}^{(1)} \ldots \boldsymbol{O}^{(K)}$. For example, for a path finding task on a graph, the process can be decomposed into K steps. At each step, we determine a single waypoint, and use this waypoint as the starting point for the next step. After the prediction of K steps, the complete path is obtained. For step K, the network input is the state output $\mathcal{X}^{(k)} = [x_1^{(k)}; \ldots; \boldsymbol{x}_{|\mathcal{V}|}^{(k)}]^\top$ of step $K-1$. Two initial models, $\mathcal{F}_O^{(k)}$ and $\mathcal{F}_X^{(k)}$, are used. $\mathcal{F}_O^{(k)}$ is used to predict the output of the current step. $\mathcal{F}_X^{(k)}$ is used to get the updated node states $\mathcal{X}^{(k+1)}$. The authors find that sharing the propagation layer in $\mathcal{F}_O^{(k)}$ and $\mathcal{F}_X^{(k)}$ can speed up inference with only a slight performance penalty. If a task cannot provide the annotation information of the intermediate step, the network can only use the supervision information of the last step for back propagation.

3.3.2 *Graph Convolutional Network*

In Chap. 1, we introduce the development history of Graph Convolutional Networks, and in this section we will introduce a typical graph convolutional network GraphSAGE [8]. The GraphSAGE is a further extension of GCN [9]. The following three formulas are the main parameter update process of GraphSAGE. $\mathbf{h}_u^k$ indicates the hidden layer state of layer k node u, and AGGREGATE_k is the aggregation operation of layer k. $\mathbf{W}^k$ is the learnable weight parameter of layer k. σ is a nonlinear activation function.

$$\mathbf{h}_{\mathcal{N}(u)}^k \leftarrow \text{AGGREGATE}_k(\{\mathbf{h}_v^{k-1}, \forall v \in \mathcal{N}(u)\}), \tag{3.28}$$

$$\mathbf{h}_u^k \leftarrow \sigma\left(\mathbf{W}^k \cdot \text{CONCAT}(\mathbf{h}_u^{k-1}, \mathbf{h}_{\mathcal{N}(u)}^k)\right), \tag{3.29}$$

$$\mathbf{h}_u^k \leftarrow \mathbf{h}_u^k / \|\mathbf{h}_u^k\|_2, \forall u \in V, \tag{3.30}$$

Equation (3.28) illustrates the aggregation process of neighbor node information. Equation (3.29) represents the process of updating the node state. The neighbor nodes in Eq. (3.28) are obtained by uniform sampling. The number of neighbor nodes in the same layer is fixed, and the number of neighbor nodes in different layers can be different. When the number of samples is greater than the degree of the node, the samples with replacement are used.

The aggregate function in Eq. (3.28) should be independent of the node order. The authors use three types of aggregators: Mean Aggregator, LSTM Aggregator, and Pooling Aggregator. Equations (3.28) and (3.29) can be replaced by the average aggregator as shown below. The average aggregator can be thought of as an approximation of the spectral convolution [9].

$$\mathbf{h}_u^k \leftarrow \sigma(\mathbf{W} \cdot \mathrm{MEANN}(\{\mathbf{h}_u^{k-1}\} \cup \{\mathbf{h}_v^{k-1}, \forall v \in \mathcal{N}(u)\}), \tag{3.31}$$

where concatenation operation is not used in AGGREGATE^{mean}. LSTMs have stronger expressive capability and are used in AGGREGATE^{lstm}. Before being input into AGGREGATE^{lstm}, the order of the nodes is randomly shuffled. The pooling aggregator AGGREGATE^{pool} employs a linear layer and a nonlinear activation function, as shown in the following equation.

$$\mathrm{AGGREGATE}_k^{\mathrm{pool}} = \max(\{\sigma\left(\mathbf{W}_{\mathrm{pool}}\mathbf{h}_{v_i}^k + \mathbf{b}\right), \forall v_i \in \mathcal{N}(u)\}), \tag{3.32}$$

where the elementwise max-pooling operator is remarked by max. Important information in the neighbor nodes is captured. The linear layer can be replaced with a multilayer perceptron. The max-pooling operator in Eq. (3.32) can also be replaced by the elementwise mean-pooling operator.

Stochastic gradient descent is adopted for network training. The authors first give the loss function in the unsupervised setting, as shown in Eq. (3.33). $J_G(\cdot)$ makes the nodes far away have obvious differences, and the neighborhood nodes have similar features. v co-occurs on a fixed length random walk path starting from node u. $\mathbf{z}_u$, $u \in V$ represents network output. The negative sampling distribution is denoted by $P_n(\cdot)$. The negative sample number is indicated by Q.

$$J_G(\mathbf{z}_u) = -\log\left(\sigma(\mathbf{z}_u^\top \mathbf{z}_v)\right) - Q \cdot \mathbb{E}_{v_n \sim P_n(v)} \log\left(\sigma(-\mathbf{z}_u^\top \mathbf{z}_{v_n})\right). \tag{3.33}$$

In a supervised setting, task-specific loss functions such as cross-entropy loss, mean square error, etc., can be used.

3.3.3 Graph Attention Networks

Graph Attention Networks (GATs) use attention mechanisms to aggregate information about node neighbors. Using the attention mechanism, aggregate weights of different neighbor nodes can be learned automatically when the vector representation of neighbor nodes is aggregated. The vector representation of nodes aggregate computation can also be performed in parallel between nodes. The parameters to be learned are fixed and independent of the size of the graph. The weights in GAT can be learned automatically, which also means that it has strong generalization ability and can be applied well to unknown nodes.

However, existing graph neural networks, such as the graph convolutional network MPNN [10], the attention network GAT [11] and their variants [12, 13], do not capture the interaction between neighbor nodes. To solve this problem, Wang et al. [14] introduce partial information decomposition [15] into graph neural networks and propose Graph Ordering Attention (GOAT). The mutual information $I(\mathbf{h}_u; \boldsymbol{H}_{\overline{\mathcal{N}}(u)})$ between nodes u and $\overline{\mathcal{N}}(u)$ consists of three parts: unique information U, redundant information R, and synergistic information S. $\mathbf{h}_u$ represents the hidden feature of node u. $\overline{\mathcal{N}}(u) = \mathcal{N}(u) \cup \{u\}$ and $\boldsymbol{H}_{\overline{\mathcal{N}}(u)} = [\mathbf{h}_{v_1}, \ldots, \mathbf{h}_{v_{|\overline{\mathcal{N}}(u)|}}]$. The information owned by one neighbor node but not by other neighbors is called unique information U. The information existing in two or more neighbor nodes is called redundant information R. The synergistic information S exists in the interaction between neighbor nodes. For the contribution coefficient c_{uv}, $c_{uv} = f(\boldsymbol{S}, \boldsymbol{H}_{\overline{\mathcal{N}}(u)})$. Therefore, c_{uv} does not only consider the hidden features of node u and v.

A GOAT layer can be divided into two parts, the ordering part and the sequence modeling part. As shown in the following equation, the ordering part OrderingPart$(\cdot)$ rearranges the feature set of neighboring nodes $\{\mathbf{h}_1, \ldots, \mathbf{h}_P\}$ into an ordered feature set, where $P = \left|\overline{\mathcal{N}}(u)\right|$. $\pi(\cdot)$ denotes the permutation function.

$$[\mathbf{h}_{\pi(1)}, \ldots, \mathbf{h}_{\pi(P)}] = \text{OrderingPart}(\{\mathbf{h}_1, \ldots, \mathbf{h}_P\}). \tag{3.34}$$

Specifically, the authors employ self-attention to compute the attention score a_{ij} between nodes i and j as shown in Eq. (3.35),

$$a_{ij} = \text{LeakyReLU}\left(\mathbf{w}_2^T [\mathbf{W}_1 \mathbf{h}_i \| \mathbf{W}_1 \mathbf{h}_j]\right), \tag{3.35}$$

where $\mathbf{W}_1 \in \mathbb{R}^{d \times d}$ and $\mathbf{w}_2 \in \mathbb{R}^{2d}$ are learnable parameters. Then, the attention scores are sorted in ascending order to obtain the permutation π as shown in Eq. (3.36), the attention coefficients are normalized by the softmax operation to construct the ranked feature set as shown in Eq. (3.37).

$$a_{i\pi(1)}, \ldots, a_{i\pi(P)} = sort\,(a_{i1}, \ldots, a_{iP}), \tag{3.36}$$

$$\begin{aligned}\mathbf{h}_{sorted(i)} = \Bigg[& \frac{e^{a_{i\pi(1)}}}{\sum_{j=1}^{Q} e^{a_{i\pi(j)}}} \mathbf{W}_1 \mathbf{h}_{\pi(1)}, \ldots, \\ & \frac{e^{a_{i\pi(Q)}}}{\sum_{j=1}^{Q} e^{a_{i\pi(j)}}} \mathbf{W}_1 \mathbf{h}_{\pi(P)} \Bigg].\end{aligned} \tag{3.37}$$

In the sequence modeling part, the bidirectional LSTM is used to update the node states as shown in Eq. (3.38). d_O represents the dimension of output node features. Synergistic information is recognized by the memory unit in the module and redundant information can be discarded by the forgetting gate. Neighboring nodes before and after node j, i.e., all other neighboring nodes, are taken into account while calculating the contribution coefficient c_{ij} of node j.

$$\mathbf{h}_i^{new} = \text{LSTM}\left(\mathbf{h}_{sorted(i)}\right) \in \mathbb{R}^{d_O}. \tag{3.38}$$

Here all other RNNs can be used. The set of neighbor nodes of different nodes can also be processed in parallel by the RNNs. The authors introduce multi-head attention into the GOAT layer. One neighbor node order is not robust enough. Multiple neighbor node orders can be captured by multi-head attention as shown in Eq. (3.39), where K denotes the number of attention heads.

$$a_{ij}^k = a^k(\mathbf{W_1^k}\mathbf{h}_i, \mathbf{W_1^k}\mathbf{h}_j), \tag{3.39}$$

$$\mathbf{h}_i^{new} = \prod_{k=1}^{K} \text{LSTM}^k\left(\mathbf{h}_{sorted(i)}^k\right). \tag{3.40}$$

The differently ordered neighbor node feature sets $\mathbf{h}_{sorted(i)}^k$ for $k \in \{1, \ldots, K\}$ are processed by the bidirectional LSTM and then concatenated to obtain the updated node feature $\mathbf{h}_i^{new}$ as shown in Eq. (3.40). The GOAT layers can be stacked or combined with other graph neural network layers.

3.4 Training Methods of Graph Neural Networks

We first describe the training method of the original GNN [2]. In the original GNN, the authors give the loss function as shown in Eq. (3.41). $G_i = (V_i, E_i) \in \mathcal{G}$ is a graph data in the training set $\mathcal{G}$. $\boldsymbol{w}$ denotes the network parameters. $\varphi_{\boldsymbol{w}}$ denotes the whole network model, which consists of a local transfer function $F_{\boldsymbol{w}}$ and a local output function $G_{\boldsymbol{w}}$. $\boldsymbol{y}_{i,j}$ is the true label of the j-th node in the i-th graph data. p denotes the number of graphs in the training set and q_i denotes the number of nodes in the i-th graph.

$$e_{\boldsymbol{w}} = \sum_{i=1}^{p}\sum_{j=1}^{q_i}(\boldsymbol{y}_{i,j} - \varphi_{\boldsymbol{w}}(G_i))^2. \tag{3.41}$$

The forward process is an iterative updating process, and finally the node state reaches the stabilization point $\boldsymbol{x}$. According to the Almeida–Pineda method [16, 17], the intermediate variable, $z(t)$, can be calculated by Eq. (3.42), where $\boldsymbol{I}$ denotes the prior information of nodes and edges, and $\boldsymbol{l}_V$ denotes the prior information of nodes. By making $\boldsymbol{A} = \frac{\partial F_w}{\partial \boldsymbol{x}}(\boldsymbol{x}, \boldsymbol{l})$ and $\boldsymbol{b} = \frac{\partial e_w}{\partial \boldsymbol{o}} \cdot \frac{\partial G_w}{\partial \boldsymbol{x}}(\boldsymbol{x}, \boldsymbol{l}_V)$, Eq. (3.43) can be obtained.

$$z(t) = z(t+1) \cdot \frac{\partial F_w}{\partial \boldsymbol{x}}(\boldsymbol{x}, \boldsymbol{l}) + \frac{\partial e_w}{\partial \boldsymbol{o}} \cdot \frac{\partial G_w}{\partial \boldsymbol{x}}(\boldsymbol{x}, \boldsymbol{l}_V), \tag{3.42}$$

$$z(t) = z(t+1) \cdot \boldsymbol{A} + \boldsymbol{b}. \tag{3.43}$$

The gradient can be calculated by Eq. (3.44). By making $\boldsymbol{c} = \frac{\partial e_w}{\partial \boldsymbol{o}} \cdot \frac{\partial G_w}{\partial \boldsymbol{w}}(\boldsymbol{x}, \boldsymbol{l}_V)$ and $\boldsymbol{d} = z(t) \cdot \frac{\partial F_w}{\partial \boldsymbol{w}}(\boldsymbol{x}, \boldsymbol{l})$, Eq. (3.45) is obtained.

$$\frac{\partial e_w}{\partial \boldsymbol{w}} = \frac{\partial e_w}{\partial \boldsymbol{o}} \cdot \frac{\partial G_w}{\partial \boldsymbol{w}}(\boldsymbol{x}, \boldsymbol{l}_N) + z \cdot \frac{\partial F_w}{\partial \boldsymbol{w}}(\boldsymbol{x}, \boldsymbol{l}), \tag{3.44}$$

$$\frac{\partial e w}{\partial \boldsymbol{w}} = \boldsymbol{c} + \boldsymbol{d}. \tag{3.45}$$

The whole procedure of gradient computation is given in Table 3.1. A detailed proof is given in the literature [2]. Then, the gradient descent strategy is used.

As to further improve the training efficiency, Gallicchio et al. propose graph echo state networks (GraphESN) [18]. The model is divided into three parts: the reservoir (an encoder), the state mapping function (a mapping layer), and the readout (a output layer). Among them, the mapping layer is used for graph level tasks. The authors believe that contractivity plays a more important role compared to architectural details. Therefore, the encoder of the GraphESN does not need to be trained,

Table 3.1 The process of gradient computation

```
BACKWARD(x, w)
    o = G_w(x, l_V);
    A = ∂F_w/∂x (x, l);
    b = ∂e_w/∂o · ∂G_w/∂x (x, l_V);
    initialize z(0), t=0;
    repeat
        z(t) = z(t+1) · A + b;
        t=t − 1;
    until ||z(t − 1) − z(t)|| ≤ ε_b;
    c = ∂e_w/∂o · ∂G_w/∂w (x, l_V);
    d = z(t) · ∂F_w/∂w (x, l);
    ∂ew/∂w = c + d;
    return ∂e_w/∂w;
end
```

and just initialize the encoder randomly after setting the value range and contraction factor. The mapping layer also has no learnable parameters. Only the output layer needs training.

In the simple case, only one linear layer is used in the output part $g_{out}(\cdot)$, as shown in Eq. (3.46). $\mathbf{y}(v) \in \mathbb{R}^{N_O}$ denotes the output for the node v. $\mathbf{W}_{out} \in \mathbb{R}^{N_O \times N_R}$ denotes the weight matrix. $\mathbf{x}(v)$ denotes the node feature. N_R and N_O denote the node feature dimension and the dimension of the output, respectively. For the graph level task, a mapping layer $\mathcal{X}(\cdot)$ is added as shown in Eq. (3.47). $\mathbf{g}$ denotes the whole input graph.

$$\mathbf{y}(v) = g_{out}(\mathbf{x}(v)) = \mathbf{W}_{out}\mathbf{x}(v), \tag{3.46}$$

$$\mathbf{y}(\mathbf{g}) = g_{out}(\mathcal{X}(\mathbf{x}(\mathbf{g}))) = \mathbf{W}_{out}\mathcal{X}(\mathbf{x}(\mathbf{g})). \tag{3.47}$$

The output layer is trained by linear regression as shown in Eq. (3.48). $\mathbf{X} \in \mathbb{R}^{N_R \times P}$ denotes the state matrix. $\bar{\mathbf{Y}} \in \mathbb{R}^{N_O \times P}$ denotes the ground-truth labels. P remarks the number of the nodes. The authors solve Eq. (3.48) by pseudo-inversion.

$$\|\mathbf{W}_{out}\mathbf{X} - \bar{\mathbf{Y}}\|_2^2. \tag{3.48}$$

Dai et al. further extend the training algorithm by introducing random sampling of nodes, called SSE (Stochastic Steady-state Embedding) [19]. The SSE divides the model into an iterative function and a link function. Both the iterative and link functions use a two-layer neural network. The loss function is shown in Eq. (3.49), where $\{W_i, V_i\}_{i=1}^2$ represents the parameters of the iterative and link functions. V^y denotes a subset of nodes. f_v^* denotes the ground-truth of a node. $\mathcal{T}_\Theta$ denotes the iterative function. g remarks the link function. $\widehat{\mathbf{h}}_v$ denotes the features of node v.

$$\begin{aligned} \min_{\{W_i, V_i\}_{i=1}^2} \mathcal{L}\left(\{W_i, V_i\}_{i=1}^2\right) &:= \frac{1}{|V^y|} \sum_{v \in V^y} \ell(f_v^*, g(\widehat{\mathbf{h}}_v)) \\ \text{s.t. } \widehat{\mathbf{h}}_v &= \mathcal{T}_\Theta\left[\{\widehat{\mathbf{h}}_u\}_{u \in \mathcal{N}(v)}\right], \forall v \in V. \end{aligned} \tag{3.49}$$

SSE views the learning process as policy iteration in reinforcement learning. The learning process contains node update steps and weight update steps. In the node update step, a subset $\widetilde{V}$ is first sampled from the node set and then the node features are updated using the moving average shown in Eq. (3.50).

$$\widehat{\mathbf{h}}_{v_i} \leftarrow (1-\alpha)\widehat{\mathbf{h}}_{v_i} + \alpha\mathcal{T}_\Theta\left[\{\widehat{\mathbf{h}}_u\}_{u \in \mathcal{N}(v_i)}\right], \forall v_i \in \widetilde{V}, \tag{3.50}$$

$$
\begin{aligned}
\frac{\partial \mathcal{L}}{\partial W_i} &= \widehat{\mathbb{E}}\left[\frac{\partial \ell\left(f_v^*, g\left(\widehat{\mathbf{h}}_v^k\right)\right)}{\partial g\left(\widehat{\mathbf{h}}_v^k\right)} \frac{\partial g\left(\widehat{\mathbf{h}}_v^k\right)}{\partial W_i}\right] \\
\frac{\partial \mathcal{L}}{\partial V_i} &= \widehat{\mathbb{E}}\left[\frac{\partial \ell\left(f_v^*, g\left(\widehat{\mathbf{h}}_v^k\right)\right)}{\partial \widehat{\mathbf{h}}_v^k} \frac{\partial \mathcal{T}_\Theta\left[\{\widehat{\mathbf{h}}_u^k\}_{u \in \mathcal{N}(v)}\right]}{\partial V_i}\right].
\end{aligned}
\tag{3.51}
$$

In the weight update step, the gradient is computed as shown in Eq. (3.51). k denotes the number of rounds of stochastic optimization. $\widehat{\mathbb{E}}[\cdot]$ denotes the expectation taken over the labeled set of nodes. Data augmentation, weakly supervised training and unsupervised training on graphs are also important directions of interest to researchers.

In this chapter, we provide an introduction to the basics of using graph neural networks. We first introduce the general framework for graph neural networks, which can be divided into three types of layers: feature transformation layers, activation layers, and pooling layers. Operations such as spectral graph convolution, spatial graph convolution, and graph attention can all be used to implement feature transformation layers. Next, we describe the typical operations and representative methods of graph neural networks, covering spectral graph convolution, spatial graph convolution, graph attention, and graph pooling. Regarding graph neural network architectures, we present notable methods for vanilla graph neural networks, graph convolutional networks, and graph attention networks, providing a comprehensive understanding of the various types of graph neural networks. Finally, we summarize the training methods for graph neural networks. Initially, the Almeida–Pineda method [16, 17] methods are used, followed by node sampling methods such as random node sampling. More recently, efficient training methods have been proposed, including data augmentation, weakly supervised learning, and unsupervised learning on graphs.

References

1. Gori, M., Monfardini, G., Scarselli, F.: A new model for learning in graph domains. In: IEEE International Joint Conference on Neural Networks, vol. 2, pp. 729–734 (2005)
2. Scarselli, F., Gori, M., Tsoi, A.C., Hagenbuchner, M., Monfardini, G.: The graph neural network model. IEEE Transactions on Neural Networks 20(1), 61–80 (2009)
3. Gao, H., Ji, S.: Graph u-nets. IEEE Transactions on Pattern Analysis and Machine Intelligence 44(9), 4948–4960 (2022)
4. Bianchi, F.M., Grattarola, D., Alippi, C.: Spectral clustering with graph neural networks for graph pooling. In: International Conference on Machine Learning, vol. 119, pp. 874–883 (2020)
5. Stergiou, A., Poppe, R., Kalliatakis, G.: Refining activation downsampling with softpool. In: IEEE International Conference on Computer Vision, pp. 10337–10346 (2021)
6. Grattarola, D., Zambon, D., Bianchi, F.M., Alippi, C.: Understanding pooling in graph neural networks. IEEE Transactions on Neural Networks and Learning Systems 35(2), 2708–2718 (2024)
7. Li, Y., Tarlow, D., Brockschmidt, M., Zemel, R.S.: Gated graph sequence neural networks. In: Proceedings of the International Conference on Learning Representations (2016)

8. Hamilton, W.L., Ying, Z., Leskovec, J.: Inductive representation learning on large graphs. In: Advances in Neural Information Processing Systems, pp. 1024–1034 (2017)
9. Kipf, T.N., Welling, M.: Semi-supervised classification with graph convolutional networks. In: International Conference on Learning Representations (2017)
10. Gilmer, J., Schoenholz, S.S., Riley, P.F., Vinyals, O., Dahl, G.E.: Neural message passing for quantum chemistry. In: International Conference on Machine Learning, vol. 70, pp. 1263–1272 (2017)
11. Velickovic, P., Cucurull, G., Casanova, A., Romero, A., Liò, P., Bengio, Y.: Graph attention networks. In: International Conference on Learning Representations (2017)
12. Brody, S., Alon, U., Yahav, E.: How attentive are graph attention networks? In: International Conference on Learning Representations (2022)
13. Zhang, S., Xie, L.: Improving attention mechanism in graph neural networks via cardinality preservation. In: International Joint Conference on Artificial Intelligence, pp. 1395–1402 (2020)
14. Chatzianastasis, M., Lutzeyer, J.F., Dasoulas, G., Vazirgiannis, M.: Graph ordering attention networks. In: In:Proceedings of the AAAI Conference on Artificial Intelligence, pp. 7006–7014 (2023)
15. Williams, P.L., Beer, R.D.: Nonnegative decomposition of multivariate information. arXiv preprint 1004.2515 (2010)
16. Almeida, L.B.: A learning rule for asynchronous perceptrons with feedback in a combinatorial environment. In: IEEE International Joint Conference on Neural Networks, vol. 2, pp. 609–618 (1987)
17. Pineda, F.J.: Generalization of back-propagation to recurrent neural networks. Physical Review Letters 59(19), 2229–2232 (1987)
18. Gallicchio, C., Micheli, A.: Graph echo state networks. In: International Joint Conference on Neural Networks, pp. 1–8 (2010)
19. Dai, H., Kozareva, Z., Dai, B., Smola, A.J., Song, L.: Learning steady-states of iterative algorithms over graphs. In: Proceedings of the International Conference on Machine Learning, vol. 80, pp. 1114–1122 (2018)

Chapter 4
Graph Convolutional Network for Scene Parsing

Semantic analysis of scene is one of the focuses in computer vision research, and scene parsing is a key technology to realize semantic segmentation and recognition of image. In some studies, researchers concentrated on image-level scene classification or attribute recognition, who integrated hand-craft image features and classification models, such as support vector machine, decision tree and multi-layer perceptron, to build scene representation and classification model. These methods sometimes were not robust for real-world applications, because the scene image contained considerable objects and structure information, which caused huge intra-class difference or inter-class similarity.

4.1 Introduction

Scene parsing is a more complex task than the human parsing and the video object segmentation, and it is also one of the hotspots in computer vision research. It requires the algorithm model to realize pixel-level scene object segmentation and recognition, which is not only a multi-label problem, but also needs to match or locate labels to specific image pixels. And it is also known as semantic segmentation. This needs to annotated scene image containing a large number of objects more densely and intensively. For example, given an indoor scene image of the living room, not only need to indicate object such as a sofa, a desk and a TV, and segmentation background region, such as the floor, wall and ceiling, also need to mark the chair seat or finer-grained components such as handle of the cup. In order to achieve pixel-level segmentation and recognition, the algorithm model would have to learn fine scene features from the original data.

W. Liu et al., *Graph Neural Network Methods and Applications in Scene Understanding*,
https://doi.org/10.1007/978-981-97-9933-6_4

In the past ten years, since the Fully Convolutional Network (FCN) [1] was proposed, Deep Convolutional Neural Network (DCNN) methods have brought great progress to scene parsing. Some typical deep models include FCN, DeepLab [2], Pyramid Scene Parsing Network (PSPNet) [3], SegNet [4] and so on. The convolutional architecture has provided powerful feature learning capacity and classification performance, making it possible to achieve coarse-grained scene parsing. Since the deep learning models focus on low-level pixel information and usually ignore structure information, it is difficult to realize the pixel-level semantic analysis of scene objects, especially the multi-scale object recognition and segmentation in the scene image. Additionly, it is hard to balance the feature learning weight of small and large objects, because the larger objects can provide more local pixel information, while the smaller objects contain few raw pixel and have to rely on neighbor structure information to enhance local feature. The deep learning model would be more inclined to learn features of large objects, resulting in a decrease in parsing performance for small objects. In recent years, Graph Convolutional Network (GCN) model has been adopted to solve graph data learning for both structure information [5] and low-level information [6] by graph convolution operation, which could help to deal with multi-scale objects. For convenience, symbols used in the ensuing presentation are summarized in Table 4.1.

Table 4.1 Symbols used in this article

Symbol	Description
$\mathbb{R}$	Set of real numbers
$\mathbb{R}^n$	Set of n-dimensional vectors
$\mathbb{R}^{m\times n}$	Set of $m \times n$-dimensional matrices
I_n	n-dimensional identity matrix
G	Graph model G
$\mathcal{V}$	Node set of a graph model
$\mathcal{E}$	Edge set of a graph model
$\mathcal{N}$	Neighbor set of a node
Λ	Adjacency matrix of a graph model
$\mathbb{C}$	The number of channel of a node feature
$\mathbb{L}$	The number of spatial level of a graph model
$\mathbb{N}$	The number of node of a graph model
K	The number of neighbor level of a node
H	Global hidden state matrix of a graph network model
O	Global output matrix of a graph network model
μ_{gain}	Feature gain parameter
μ_{gamma}	Feature nonlinearity parameter
μ_{offset}	Feature offset parameter
$*_G$	Spectral graph convolution
$\odot$	Element-wise multiplication

4.2 Graph Construction of Scene Image

Graph construction projects the original scene image into topological space to obtain scene graph data, and its essence is to transform RGB image data into graph structure data. Scene image is no longer represented by pixels matrix, but by interconnected graph nodes, and the connection of two nodes is edge. Therefore, the construction of graph structure data is the process of mapping original image pixels to graph nodes and edges.

In order to realize the transformation of the graph structure data, a hierarchical scene graph data construction method based on multi-granularity mesh is proposed in this study. Firstly, as shown in Fig. 4.1, there are different meshes on the original scene image and the grid size is decreasing from left to right, then a grid will be a graph node. Obviously, the minimum granularity is taking an original pixel as a graph node, and the construction is based on the semantic relationship of pixels, which is usually given by the annotation of the dataset. If a node has same semantic label with one of its neighbor nodes, these two nodes will be connected by a spatial edge. Then, based on multi-granularity mesh, the spatial graph of scene image is constructed by establishing spatial edges through four neighbors of grid. As shown in Fig. 4.1a, solid lines connecting two grids (nodes) are spatial edges of graph. Then the spatial graphs constructed under different granularity are superimposed to establish spatial hierarchical edges. Finally the hierarchical scene graph data is obtained, as shown in Fig. 4.2. The dashed line connecting two nodes on the right is the spatial hierarchical edge of the hierarchical scene graph.

Each scene image is divided into hierarchical grids with different granularity, then spatial edges and spatial hierarchical edges are established based on four neighbors to form the graph structure, namely, hierarchical scene graph. On this basis, each grid is further extracted by pre-defined feature engineering algorithm to obtain the attribute information carried by feature vector as each node, so as to construct scene graph feature data based on node attribute information.

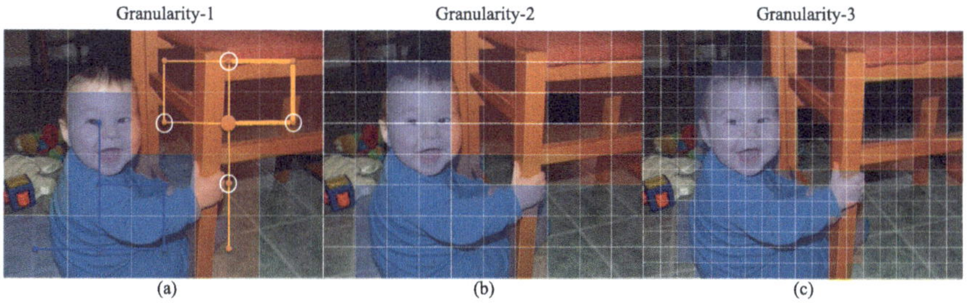

Fig. 4.1 Different grid size for scene image by multi-granularity mesh. Nodes with white circle are four neighbors of the center node

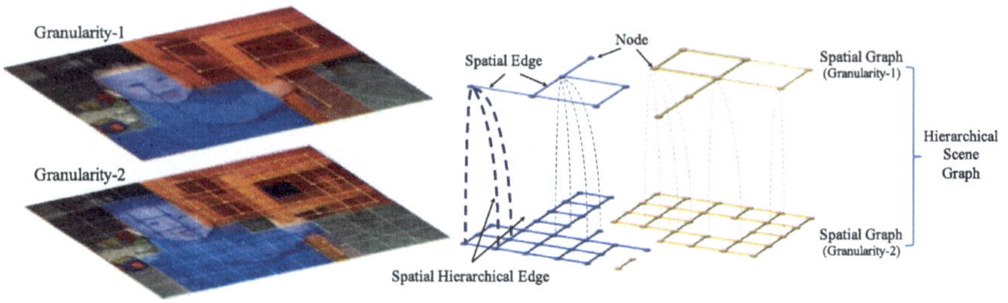

Fig. 4.2 Graph construction based on four neighbors of grid. In the hierarchical scene graph, each node is a grid of the original scene image

The formal representation is given below. Firstly, the hierarchical scene graph data can be expressed as the following:

$$G = (\mathcal{V}, \mathcal{E}_{\mathrm{S}}, \mathcal{E}_{\mathrm{SH}}), \tag{4.1}$$

where $\mathcal{V}$ is the set of graph nodes, $\mathcal{E}_{\mathrm{S}}$ is the set of spatial edges, $\mathcal{E}_{\mathrm{SH}}$ is the set of spatial hierarchical edges. Its adjacency matrix can be expressed as:

$$\Lambda = (\Lambda_{\mathrm{S}}, \Lambda_{\mathrm{SH}}), \tag{4.2}$$

where Λ_{S} is the adjacency matrix based on spatial edge, and Λ_{SH} is the adjacency matrix based on spatial hierarchical edge. Element in Λ is discrete, denoted as $e_{u,v} \in \{0, 1\}$, that if there is an edge between node v and v, it will have $e_{u,v} = 1$, otherwise $e_{u,v} = 0$. By integrating the Eqs. (4.1) and (4.2), the hierarchical scene graph data structure is shown in the following:

$$G = (\mathcal{V}, \mathcal{E}_{\mathrm{S}}, \mathcal{E}_{\mathrm{SH}};\ \Lambda_{\mathrm{S}}, \Lambda_{\mathrm{SH}}). \tag{4.3}$$

Then, after feature extraction of the original scene image grids, the scene graph feature data for model input can be expressed in the form of the set shown in Eq. (4.4):

$$X = \{\boldsymbol{x}_{n,l} | \boldsymbol{x}_{n,l} \in \mathbb{R}^{\mathbb{C}}, v_{n,l} \in \mathcal{V}, n \in [1, \mathbb{N}_l], l \in [1, \mathbb{L}]\}, \tag{4.4}$$

where $\boldsymbol{x}_{n,l}$ is the attribute information carried by graph node $v_{n,l}$. In the following a self-supervised feature fusion-based graph convolutional network (SFGCN) model building, the attribute information is also called the feature of graph node, whose dimension (the number of data channel) is $\mathbb{C}$. Then, the number of spatial levels is $\mathbb{L}$ and the number of nodes for the lth level is $\mathbb{N}_l$. Let $\mathbb{N} = \max(\mathbb{N}_l)$, it will have $X \in \mathbb{R}^{\mathbb{N}\times\mathbb{L}\times\mathbb{C}}$. Furthermore, let $\mathbb{N} = \sum_{l=1}^{\mathbb{L}} \mathbb{N}_l$, X can be simplified as matrix, denoted as $X \in \mathbb{R}^{\mathbb{N}\times\mathbb{C}}$. The following uses $\mathbb{N}$ to refer to the total number of nodes in the set $\mathcal{V}$. The node features can also be obtained by the learnable linear transformation $f(\mathbf{x}, \boldsymbol{w})$, as shown in the Eq. (4.5):

$$x_{n,l} = f(\mathbf{x}, \boldsymbol{w}) = \boldsymbol{w}_{n,l}^{\mathrm{T}} \mathbf{x}_{n,l} + b, \tag{4.5}$$

where $\mathbf{x}$ represents the original image data of the node $v_{n,L}$, $\boldsymbol{w}$ represents the parameterized linear weight, and $\boldsymbol{b}$ represents the bias. For convenience, the following also uses $\boldsymbol{x}_v$ and $\mathbf{x}_v$ as attributes and original data of node v respectively, equivalent with $\boldsymbol{x}_{n,l}$ and $\mathbf{x}_{n,l}$.

Through the above steps, the raw input data can be converted into hierarchical scene graph feature data, and as the input module adding to the SFGCN model. This module is a kind of normalized processing for original image data and is able to format pixel data into graph data. After this, each node in the graph structure has individual attribute information (feature). The graph structure will be shared among all layers of the model so as to train the weight parameters of the graph convolution layer.

4.3 Overview of the GCN Architecture

In this study, a graph convolution network model based on self-supervised feature fusion is proposed for scene parsing. Based on basic GCN which uses the plain spatial graph convolution for feature learning, this method proposes graph attention module based on self-supervised feature fusion, improves spatial graph convolution by K-level neighbor, and designs graph pooling module for scene domain adaptivity, etc., to realize pixel-level semantic label parsing of multi-scale scene object.

Specifically, the problem of multi-scale object in scene parsing is studied from the perspective of graph model. Firstly, hierarchical meshing is used to construct scene graph feature data, and then a graph convolutional network model based on graph attention module with self-supervised feature fusion is constructed, aiming at the difficulty of parsing local semantic labels. By combining spectral graph convolution and attention mechanism, self-supervised information is added to model weight training to enhance local feature learning. Then the K-level neighbor-based graph convolution operation is proposed to realize global feature learning. Finally, in order to maintain the intra-domain consistency of multi-domain scene images, a multi-domain adaptive scene graph pooling method is proposed to prevent the intra-class inconsistency of scene images caused by their own imaging differences. Experimental results on several large-scale datasets show that this method can effectively implement scene parsing and is superior to other advanced methods.

To build the architecture of SFGCN, by referring to the multi-layer feature extraction and neural cell-based architecture of DCNN model, this research firstly designs the cell based on plain spatial graph convolution and the Self-supervised Feature fusion graph Attention (SFA) module, then builds network model by stacking multi-layer neural cells. The tth cell is also the tth layer of the model. This architecture is called SFGCN, showing in Fig. 4.3a. Where, both the plain convolution and SFA operation combine with activation function (ReLU), realizing nonlinear graph feature transformation. As shown in Fig. 4.3b, the architecture of SFGCN-SFA is built by

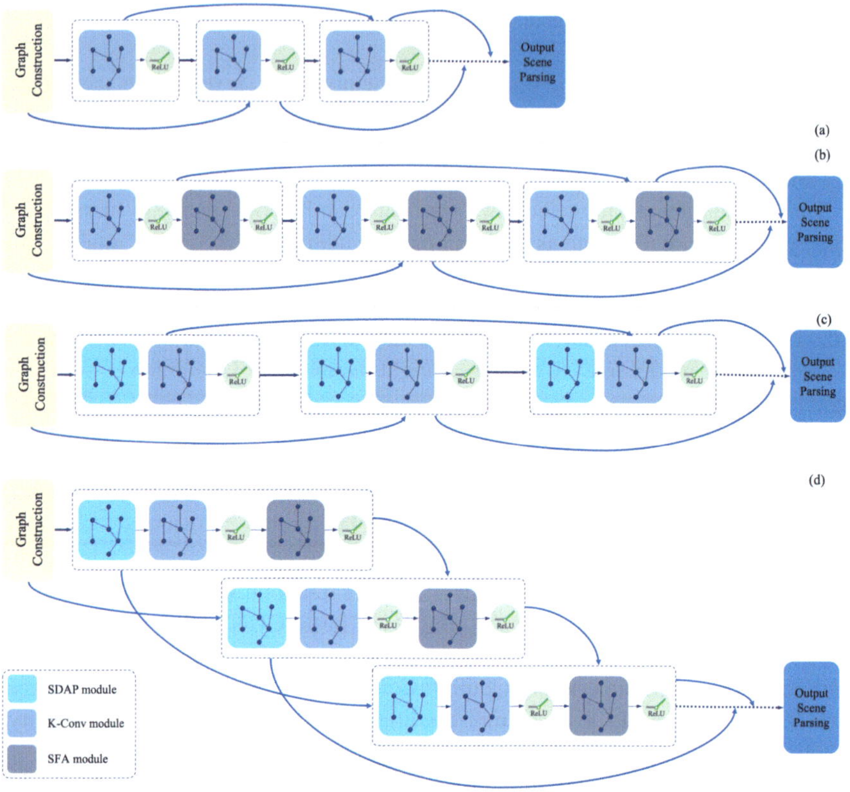

Fig. 4.3 Architecture of GCN for scene parsing. **a** SFGCN **b** SFGCN-SFA **c** SFGCN-SDAP **d** SFGCN-ALL

replacing plain spatial graph convolution by K-Conv to strengthen the feature learning ability of graph attributes and graph connections for the network model. The Scene Domain Adaptive graph Pooling (SDAP) module is added before K-Conv to build the architecture of SFGCN-SDAP, as shown in Fig. 4.3c. Finally, SFGCN-ALL is shown in Fig. 4.3d. As shown in Fig. 4.3, modules of those four architectures are different. The dashed box shows the constituent structure of a neural cell, each cell connects to previous two cells and then the chain architecture is used to construct the graph convolutional network model.

Since the SFA module is usually placed after the K-Conv module to enhance the feature learning, the next part of this section will introduce the K-level neighbor-based spatial graph convolution first, and then introduce the feature fusion and self-supervised attention mechanism of SFA. The last two parts will introduce the SDAP module and the training of SFGCN.

4.4 Spatial Graph Convolution Based on K-Level Neighbour

In the graph convolutional network model, the spatial graph convolution module has the effect of feature learning, the plain convolution-based message passing (MP) only updates the current node state by computing messages from nearest neighbors, which does not benefit nodes with too few neighbors, resulting in over-fitting of model weights. This study improves the plain convolution module and proposes K-level neighbor based spatial graph convolution operation, namely K-Conv, which lets the neighbor of neighbor pass its message to the current node. The nearest neighbor is level-1, the neighbor of the nearest neighbor is level-2, and so on. Through collecting message from multi-level neighbors, every node can obtain more information for updating model weight. According to different edge types (see Eq. 4.1), one node may have spatial neighbor connected by spatial edge or spatial hierarchical neighbor connected by spatial hierarchical edge. Both or one of the neighbors can be used for K-Conv operation.

An output state of the node v at model layer (t) is shown in the Eq. (4.6) without considering the activation function temporarily:

$$\mathbf{h}_v^{(t)} = (\boldsymbol{w}_v^{\mathrm{T}})^{(t)} \boldsymbol{x}_v + \sum_{k=1}^{\mathrm{K}} \sum_{u \in \mathcal{N}_k(v)} (\boldsymbol{\theta}_k^{\mathrm{T}})^{(t)} \mathbf{h}_u^{(t-1)}, \tag{4.6}$$

where $\mathbf{h}_v^{(0)} = 0$, and $\boldsymbol{w}_v$ indicates the attribute weight of node v, $\boldsymbol{\theta}$ indicates the edge connection weight of node v, K is a constant, which denotes the K-level neighbor of node v, then $\mathcal{N}_k(v)$ is set of level-k neighbor of node v. Its matrix form can be written as the Eq. (4.7):

$$H^{(t)} = XW^{(t)} + \sum_{k=1}^{\mathrm{K}} \Lambda H^{(t-1)} \Theta_k^{(t)}, \tag{4.7}$$

where Λ is the adjacency matrix, and Θ_k is the edge connection weight matrix of the node neighbor of the layer.

In the Eq. (4.6), it can also get $\mathbf{h}_v^{(0)}$ by Eq. (4.5), and let $[\boldsymbol{w}_v^{(0)}]_0 = 1$, $[\boldsymbol{x}_v]_0 = b$, where $[\cdot]_0$ is the 0th element of the vector. As shown in Eq. (4.8), at this time, the attributes of the original data are learned directly through spatial graph convolution operation:

$$\mathbf{h}_v^{(0)} = (\boldsymbol{w}_v^{\mathrm{T}})^{(0)} \boldsymbol{x}_v. \tag{4.8}$$

Based on message passing mechanism, convolution operation process of Eq. (4.6) and node state update process can be reformulated as message passing and state update process. The message function Message$(\cdot)$ and the update function Update$(\cdot)$ are expressed by the Eqs. (4.9) and (4.10):

$$\boldsymbol{m}_v = \sum_{k=1}^{\mathrm{K}} \sum_{u \in \mathcal{N}_k(v)} \text{Message}(\mathbf{h}_u), \tag{4.9}$$

$$\boldsymbol{x}_v^{'} = \text{Update}(\boldsymbol{x}_v, \boldsymbol{m}_v). \tag{4.10}$$

Using Eq. (4.6) can deduce a new node attributes $\boldsymbol{x}_v^{'} \in \mathbb{R}^{\mathbb{C}^{'}}$. For fixed graph structure, complete state output can be represented as $X^{'} \in \mathbb{R}^{\mathbb{N}\times\mathbb{L}\times\mathbb{C}^{'}}$. Therefore, based on the K-Conv operation, there is a feature mapping $F : X \rightarrow X^{'}$, which implement high-level feature learning of each node. And the feature will be node representation for scene parsing.

4.5 Graph Attention Module Based on Self-Supervised Feature Fusion

The K-Conv operation mainly focuses on the global feature learning of scene graph data. In order to improve the feature learning ability of the model for small-scale objects, this study further proposes a feature fusion method based on spectral graph convolution to realize global and local feature fusion. Besides, in order to learn the real connection relation of nodes in scene graph feature data, a graph attention module based on self-supervised information, namely SFA, is constructed to enhance the learning of connection relations between local nodes for the global model training.

Specifically, as shown in the Eq. (4.6), K-Conv operation mainly focuses on the feature learning of global node attributes, which requires the connection relation of graph nodes given by adjacency matrix Λ as a known condition to enable node attributes in the neighbor to carry out message along the edge. However, only a static Λ is constructed at the model input layer. For small-scale objects, there are few neighbor connections, and the model is also weak in feature learning for such objects. To solve this problem, on one hand, the multi-level neighbors of a given node are extracted to form a local graph structure, and then the feature fusion mechanism can be established through a local spectral graph convolution model. This allows the model to combine the global and local feature of nodes. On the other hand, the local spectral graph convolution model also provides self-supervised information for further global model training, and the graph attention module is adopted to enhance local connection relation learning.

4.5.1 Local Spectral Graph Convolution Model

Spectral graph convolution is an operation of graph signal processing defined on graph Laplacian matrix in the frequency domain. The Laplacian matrix of the graph can be expressed as $L = \mathrm{D} - \Lambda$, where D is the diagonal matrix of node degree and Λ is the adjacency matrix of the graph. The Laplacian spectrum is the eigenvalue of L. And then define normalized Laplacian matrix as $L = I_n - \mathrm{D}^{-\frac{1}{2}}\Lambda\mathrm{D}^{-\frac{1}{2}}$, where $\mathrm{D}_{ii} = \sum_j \Lambda_{i,j}$, L is a real semi-positive definite symmetric matrix, which can be obtained by eigenvalue decomposition (EVD) as follows.

$$L = U\bar{\Lambda}U^{\mathrm{T}}, \tag{4.11}$$

where $\bar{\Lambda}$ is the eigenvalue diagonal matrix (i.e. spectrum), and U is the eigenvector matrix.

Based on the above graph signal processing method, the feature fusion of local graph structure is realized by spectral graph convolution operation, and then local spectral graph convolution model can be established. Firstly, let $\boldsymbol{x} \in \mathbb{R}^{\mathbb{C}}$, and define the Fourier transform (Eq. (4.12)) and inverse Fourier transform (Eq. (4.13)) of the graph as follows,

$$F(\boldsymbol{x}) = U^{\mathrm{T}}\boldsymbol{x}, \tag{4.12}$$

$$F^{-1}(\hat{\boldsymbol{x}}) = U^{\mathrm{T}}\hat{\boldsymbol{x}}, \tag{4.13}$$

where U is the eigenvector matrix obtained by EVD in the Eq. (4.11). The input raw graph data is projected into an orthogonal space with base U^{T} by Fourier transform, denoted as $\boldsymbol{x} = \sum_i \hat{x}_i u_i$. Therefore, the spectral graph convolution operation for input signal $\boldsymbol{x}$ by filer g can be defined as

$$x *_G g = F^{-1}(F(x) \odot F(g)) = U(U^{\mathrm{T}}\boldsymbol{x} \odot U^{\mathrm{T}}g), \tag{4.14}$$

where $*_G$ denotes spectral graph convolution, $\odot$ is element-wise multiplication. Then let $g_\theta = diag(U^{\mathrm{T}}g)$, Eq. (4.14) can be simplified to

$$x *_G g_\theta = Ug_\theta U^{\mathrm{T}}\boldsymbol{x}, \tag{4.15}$$

where, let $g_\theta = \Theta_{i,j}^{(t)}$ and $\Theta_{i,j}^{(t)}$ is a diagonal matrix. In the local spectral graph convolution model, $\Theta_{i,j}^{(t)}$ will be a learnable parameters set. Then, the (t)th layer of spectral graph convolution can be defined as

$$H_{:,j}^{(k)} = \sigma(\sum_{i=1}^{\mathbb{C}_{(t-1)}} U\Theta_{i,j}^{(t)}U^{\mathrm{T}}H_{:,i}^{(t-1)}),\ (j = 1, 2, \ldots, \mathbb{C}_{(t)}), \tag{4.16}$$

where $H^{(0)} = X$, $\mathbb{C}_{(t)}$ and $\mathbb{C}_{(t-1)}$ are the number of output channels of (t)th and $(t-1)$th layer, $\sigma(\cdot)$ is the activation function. The eigenvalue matrix $\bar{\Lambda}$ and eigenvector matrix U in the Eq. (4.11) are sensitive to the change of graph structure, which can be learned by g_θ from the Eq. (4.15). Using this method, the edge weights are updated in backward calculation.

4.5.2 *Approximation by Chebyshev Polynomial for Spectral Graph Convolution*

In the spectral graph convolution operation, it has to compute the EVD of L repeatedly, and the time complexity of EVD is $O(n^3)$. Therefore, in order to improve the operation speed and accelerate the model training, it can use the Chebyshev polynomial to obtain the approximate g_θ, so as to reduce the time complexity to $O(m)$. Here the Chebyshev polynomial is defined recursively as Eq. (4.17):

$$\begin{aligned} T_0(\boldsymbol{x}) &= 1, \\ T_1(\boldsymbol{x}) &= x, \\ T_i(\boldsymbol{x}) &= 2\boldsymbol{x}T_{i-1}(\boldsymbol{x}) - T_{i-2}(\boldsymbol{x}). \end{aligned} \tag{4.17}$$

Furthermore, define $g_\theta = \sum_{i=0}^{\mathrm{R}} \theta_i T_i(\tilde{\Lambda})$, where $\tilde{\Lambda} = \frac{2\Lambda}{\lambda_{\max}} - I_n$ (normalized to $[-1, 1]$), it can approximate the spectral graph convolution operation by (4.18), then let $\tilde{L} = \frac{2L}{\lambda_{\max}} - I_n$, and $T_i(\tilde{L}) = UT_i(\tilde{\Lambda})U^{\mathrm{T}}$, it will have (4.19).

$$\boldsymbol{x} *_G g_\theta = U\Big(\sum_{i=0}^{\mathrm{R}} \theta_i T_i(\tilde{\Lambda})\Big)U^{\mathrm{T}}\boldsymbol{x}, \tag{4.18}$$

$$\boldsymbol{x} *_G g_\theta = \sum_{i=0}^{\mathrm{R}} \theta_i T_i(\tilde{L})\boldsymbol{x}, \tag{4.19}$$

where R is the approximation order of Chebyshev polynomial. Assume that $\lambda_{\max} = 2$, then $\tilde{L} = -\mathrm{D}^{-\frac{1}{2}}\Lambda\mathrm{D}^{-\frac{1}{2}}$, when $\mathrm{R} = 1$, the Eq. (4.20) is the first order approximation; when $\mathrm{R} = 2$, the Eq. (4.21) is the second order approximation.

$$\boldsymbol{x} *_G g_\theta = \theta_0\boldsymbol{x} - \theta_1\mathrm{D}^{-\frac{1}{2}}\Lambda\mathrm{D}^{-\frac{1}{2}}\boldsymbol{x}, \tag{4.20}$$

$$\begin{aligned} \boldsymbol{x} *_G g_\theta = {} & \theta_0\boldsymbol{x} - \theta_1\mathrm{D}^{-\frac{1}{2}}\Lambda\mathrm{D}^{-\frac{1}{2}}\boldsymbol{x} \\ & + \theta_2(2\boldsymbol{x}\mathrm{D}^{-\frac{1}{2}}\Lambda\mathrm{D}^{-\frac{1}{2}} - 1)\boldsymbol{x}, \end{aligned} \tag{4.21}$$

where assume that $\theta = \theta_0 = -\theta_1$, the first order approximation of Chebyshev polynomial can be simplified as (4.22).

$$\boldsymbol{x} *_G g_\theta = \theta(I_n + \mathrm{D}^{-\frac{1}{2}}\Lambda\mathrm{D}^{-\frac{1}{2}})\boldsymbol{x}. \tag{4.22}$$

4.5.3 Self-Supervised Feature Fusion

Through the above steps, it can realize the spectral convolution operation in the graph frequency domain based on the graph Laplacian matrix. Then with the help of Laplacian matrix that is sensitive to changes of graph connection, the model can capture local node connection relations rapidly and implement feature fusion between global and local models. On the basis of obtaining global features through K-Conv, the feature fusion mechanism based on spectral graph convolution is used to combine global and local features. Furthermore, the parametric graph frequency domain filter is used to learn high-level features that represent the connection relationship between nodes.

In the process of graph construction, the feature of node is a kind of local information to the original scene image. It is generally believed that the information of a node is determined by other neighbors. Conversely, a node also determines other neighbors. Based on this prior idea, this study builds a graph learning model for the local graph structure through the spectral graph convolution operation, so that the graph node information can propagate along the connection relation in the local graph structure, enhancing the model learning of local graph features. Moreover, the connection between nodes is also a kind of local information. The feature fusion operation based on local spectral graph convolution can capture the node connection information, and the high-level feature representation of the real connection relation between nodes can be learned to enhance the feature learning of local connection. Therefore, on the basis of spatial graph convolution for graph node feature learning, spectral graph convolution can also be used for graph connection feature learning, so as to realize the feature fusion operation based on local spectral graph convolution. In this process, the local spectral graph convolution model will be obtained, and the graph node features and connection features output by this model can provide additional self-supervised information to enhance the feature learning of subsequent network modules.

4.5.4 Graph Attention Mechanism

Due to the global sharing of the graph structure in the network model, the message passing process of the Eqs. (4.9) and (4.10) can only proceed along the predefined path of Λ. It means that the model will carry out the same efficiency of feature learning for nodes with key attributes and nodes with noises, which is not conducive to model training and tuning.

Based on the self-supervised feature fusion mechanism, the local spectral graph convolution model provides self-supervised information for global model training. Furthermore, in order to improve the efficiency of feature learning, a graph attention module can be designed by parameterized self-supervised attention mechanism, introducing certain dynamic changes to the network model to make the nodes within K-level neighbor no longer pass message with the same weight.

In this study, a weighted adjacency matrix is used to implement the self-supervised attention mechanism, and the weight of connection represents the intensity of attention. Given node v, assuming that a node u contains key attributes in its neighbor, then it should strengthen the attention of v to u, which means that there is a stronger connection between v and u than other nodes. Let $e_{u,v} \in (0, 1)$ denote the relationship, the bigger $e_{u,v}$ is, the stronger attention is. Given node feature $\boldsymbol{x}_u$ and $\boldsymbol{x}_v$, let $\Phi \in \mathbb{R}^{\mathbb{C}\times\mathbb{C}}$ be a learning parameter matrix. Then the graph attention mechanism could be defined as a mapping $a : \mathbb{R}^{\mathbb{C}} \times \mathbb{R}^{\mathbb{C}} \to \mathbb{R}$ shown in the Eq. (4.23):

$$\begin{aligned} e_{u,v} &= \text{Softmax}_{u\in\mathcal{N}(v)}(\Phi\boldsymbol{x}_u, \Phi\boldsymbol{x}_v) \\ &= \frac{\exp(||\Phi\boldsymbol{x}_v||)}{\sum_{u\in\mathcal{N}(v)}\exp(||\Phi\boldsymbol{x}_u||)}, \end{aligned} \tag{4.23}$$

where $||\cdot||$ is the vector norm. Further, the Eqs. (4.6) and (4.7) can be adjusted to the following Eqs. (4.24) and (4.25) respectively.

$$\mathbf{h}_v^{(t)} = (\boldsymbol{w}_v^{\mathrm{T}})^{(t)}\boldsymbol{x}_v + \sum_{k=1}^{\mathrm{K}}\sum_{u\in\mathcal{N}_k(v)}(\boldsymbol{\theta}_k^{\mathrm{T}})^{(t)}\mathbf{h}_u^{(t-1)}e_{u,v}, \tag{4.24}$$

$$H^{(t)} = XW^{(t)} + \sum_{k=1}^{\mathrm{K}}\Lambda_{\Phi}^{(t-1)}H^{(t-1)}\Theta_k^{(t)}. \tag{4.25}$$

This mechanism can achieve different attention intensity by strengthening or weakening local connections. It does not change the number of nodes or edges, but changes the weight of connections between nodes, which means that the value in the adjacency matrix Λ_Φ is no longer discrete, but is $e_{u,v} \in (0, 1)$ in the continuous domain. It makes the node with key attributes play a more important role in information transmission, so that the network model can learn the node more effectively. In addition, this mechanism also plays a certain inhibitory role on network connections, avoiding the simultaneous activation of too many connections in the network. On one hand, it can accelerate model training, and on the other hand, it can alleviate the problem of over-fitting.

4.6 Graph Pooling Module Based on Scene Domain Adaptive Model

In this section, we firstly analysis the characteristics of multi-domain scene images and propose a scene domain adaptive model. Then, we design a graph pooling module based on the adaptive model, namely SDAP.

4.6.1 Scene Domain Adaptive Model

By observing images in scene datasets, scene could be classified into several groups of domain, as shown in Fig. 4.4, including natural scene, outdoor (urban) scene, indoor scene, etc. In different scene domains, the image acquisition is affected by different environmental factors, such as lighting, weather, and camera imaging performance etc. This leads to the possibility of huge differences between the same object, the same category or the same scene attributes. All of this mainly reflect in the image brightness, color saturation, etc., called intra-domain difference of scene. As shown in Fig. 4.5, all images are indoor scenes, in which, (a) and (c) are under natural lighting conditions, while (b) and (d) are under indoor lighting conditions, the brightness of (b) is obviously higher than (d), and also higher than that of the former two. To solve this problem, the general method is to normalize images in the process of data preprocessing. First, we calculate the RGB mean of all images, and then subtract the mean from each image. This normalization method will treat all scene images in the same way, which will lead to the loss of intra-domain consistency of scene images, and then affect the network model to learn the features of different scene domains.

Fig. 4.4 Imaging difference among 3 classical domains of scene. **a**, **b** and **c** are natural scene. **d**, **e** and **f** are outdoor (urban) scene. **g**, **h** and **i** are indoor scene

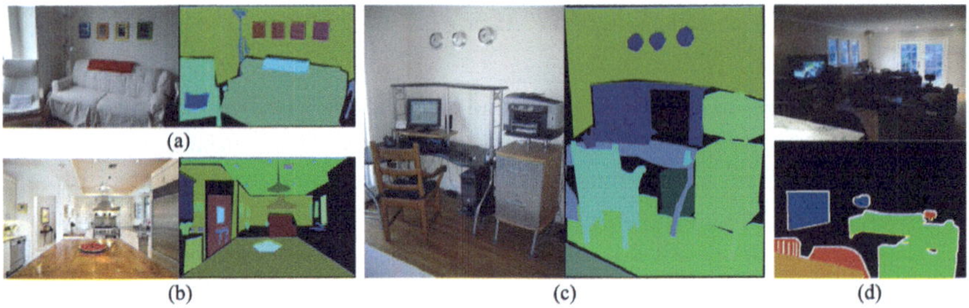

Fig. 4.5 Imaging difference of images from indoor scene domain

In order to solve the above problems, inspired by feature transformation and camera imaging parameters, the following scene domain adaptive model is proposed to perform parametric adaptive learning for the imaging differences of different scene domains. The model parameters includes feature gain parameters μ_{gain}, feature nonlinear parameter μ_{gamma} and feature offset parameter μ_{offset}. The transformation to original data for each parameter is shown in the Eq. (4.26):

$$\begin{aligned} \boldsymbol{x}_{\text{tar}} &= \mu_{\text{gain}} \cdot \boldsymbol{x}_{\text{ref}} \\ \boldsymbol{x}_{\text{tar}} &= \left(\frac{\boldsymbol{x}_{\text{ref}}}{C}\right)^{\mu_{\text{gamma}}} \cdot C \\ \boldsymbol{x}_{\text{tar}} &= \boldsymbol{x}_{\text{ref}} + \mu_{\text{offset}}, \end{aligned} \tag{4.26}$$

where $\boldsymbol{x}_{\text{ref}}$ represents the original feature of graph nodes, $\boldsymbol{x}_{\text{tar}}$ represents new feature after transformation; C is a normalized parameter, which can be $C = 2^{\text{d}} - 1$ when the input is the original pixel value, and d is the bit-width of pixel, whereas C can be set according to the channel number of feature when the input is the feature of node. By integrating all transformation parameters, it can obtain the scene domain adaptive model as Eq. (4.27):

$$\boldsymbol{x}_{\text{tar}} = \mu_{\text{gain}} \cdot \left(\frac{\boldsymbol{x}_{\text{ref}}}{C}\right)^{\mu_{\text{gamma}}} \cdot C + \mu_{\text{offset}}. \tag{4.27}$$

4.6.2 *Graph Pooling Mechanism*

Based on the adaptive model, a graph pooling module is proposed to realize the multi-domain adaptive graph pooling processing. On one hand, it can realize adaptive learning for scene domain differences, and on the other hand, it can reduce redundant information and speed up network model training. The graph pooling module mainly accomplishes two tasks, one is pooling the node features by domain adaptive learning, the other is pooling the graph structure by subgraph clustering.

Firstly, pooling feature of nodes has a direct effect on feature's dimension but has nothing to do with graph structure. Maximum pooling, denoted as Eq. (4.28), and average pooling, denoted as Eq. (4.29), can be used for down-sampling of features to reduce the learning of redundant information.

$$X_{\text{tar}} = \mu_{\text{gain}} \cdot \left(\frac{\text{pool}_{\text{max}}(X_{\text{ref}})}{C}\right)^{\mu_{\text{gamma}}} \cdot C + \mu_{\text{offset}}, \tag{4.28}$$

$$X_{\text{tar}} = \mu_{\text{gain}} \cdot \left(\frac{\text{pool}_{\text{avg}}(X_{\text{ref}})}{C}\right)^{\mu_{\text{gamma}}} \cdot C + \mu_{\text{offset}}, \tag{4.29}$$

where X_{ref} represents the original feature map. X_{tar} represents new feature map after pooling and domain adaptive transformation, pool_{max} and pool_{avg} denotes maximum graph pooling and average graph pooling respectively.

Secondly, pooling the graph structure will change the internal adjacency matrix (see Eq. (4.3)) constructed by the graph construction on scene image, which is mainly manifested by the reduction of nodes or edges, and it will obtain a new graph structure. The graph structure pooling is formulated as:

$$G^{'} = \text{pool}_G(G), \tag{4.30}$$

where, G is the original graph structure. In this chapter, $\text{pool}_G(\cdot)$ is implemented by graph pooling based on subgraph clustering. After selecting or generating cluster centers from original graph nodes and completing clustering, only cluster center nodes are retained to implement graph structure down-sampling.

The method consists of two steps. The first, node clustering is performed, known that the node set $\mathcal{V}$ of the graph structure G includes $\mathbb{N}$ nodes. Assuming that $\mathcal{V}^{'}$ is the nodes set by pooling the structure of graph $G^{'}$, which contains $\mathbb{N}^{'}$ cluster nodes. $C(v^{'})$ denotes the node set with $v^{'}$ as the cluster center, and $C(v^{'}) \subseteq \mathcal{V}$ is a subgraph of G. If $v^{'} \in \mathcal{V}$, it is a selective cluster. If $v^{'} \notin \mathcal{V}$, it is a generative cluster. Then a parameterized cluster assignment matrix can be defined as $S \in \mathbb{R}^{\mathbb{N}\times\mathbb{N}^{'}}$. The adjacency matrix of cluster nodes is shown by the Eq. (4.31).

$$\Lambda^{'} = S^{\mathrm{T}} \Lambda S, \tag{4.31}$$

where Λ is the adjacency matrix of the graph G. The second, node attribute embedding is carried out, and the node attribute set is known as $X \in \mathbb{R}^{\mathbb{N}\times\mathbb{D}}$. Define a parameterized attribute transformation matrix $Z \in \mathbb{R}^{\mathbb{N}\times\mathbb{D}}$. According to the Eq. (4.32), the node attribute set after graph pooling can be obtained by the following function:

$$X^{'} = S^{\mathrm{T}} Z, \tag{4.32}$$

where $X^{'} \in \mathbb{R}^{\mathbb{N}^{'}\times\mathbb{D}^{'}}$. This graph pooling process can be regarded as a supervised graph node classification problem. Therefore, the clustering assignment matrix S and attribute transformation matrix Z can be learned and obtained by embedding a GNN model with the adjacency matrix and node features of the original graph structure as inputs. The GNN model $\mathcal{G}$ is defined to learn graph embedding attribute and output attribute transformation matrix, and then it obtains the clustering assignment matrix by node classification based on the output attributes as follows:

$$\begin{aligned} Z &= \mathcal{G}(\hat{\Lambda}, X) \\ S &= \text{Classify}(\mathcal{G}(\hat{\Lambda}, X)), \end{aligned} \tag{4.33}$$

where Classify$(\cdot)$ represents the graph node classification function. $\mathbb{N}^{'}$ is a predefined constant in order to preserve the continuity of parameter learning before and after graph pooling and to avoid the problem of post-retraining caused by cluster node changes. GNN model $\mathcal{G}$ uses the RNN framework, defines the global transformation function F is a compression mapping, and lets $\mathcal{G}$ is the global output function. Then, there is:

$$\begin{aligned} H^{(t+1)} &= F(\hat{\Lambda}, X, H^{(t)}) \\ O &= \mathcal{G}(X, H^{(t+1)}), \end{aligned} \tag{4.34}$$

where H and O represent global implicit state and global output respectively, and (t) represents the number of GNN model layers.

4.7 Training of SFGCN Model

The original image data is transformed by hierarchical scene graph data construction method based on multi-granularity mesh, and it inputs the scene graph data into the network model. The real semantic labels of scene images are used as training labels and the model is trained by supervised learning method. In different architectures of SFGCN, the K-Conv operation is used as the global feature learning method, and combined with the SFA module and SDAP module, local feature learning and data reduction are realized respectively, so as to improve the representation ability of the model on the connection relation between nodes. Through the weighted cross entropy loss function, the trained model can output node connection prediction of scene graph data, which is the result of scene parsing, as shown in the Eq. (4.35):

$$\begin{aligned} \mathcal{L}(y, y^*) = -\sum_{v}^{\mathcal{V}} (1-\eta) y_v^* \log(y_v) \\ + \eta(1 - y_v^*)\log(1 - y_v), \end{aligned} \tag{4.35}$$

where y^* is the true label of the node, y is the predicted output of the model, and η is the loss balance weight.

4.8 Experimental Results and Discussions

Scene images usually provide little information for small objects, and it is difficult for deep models to balance the weight of feature learning between large and small objects. Hence, deep convolutional neural network faces great challenges in small-scale object segmentation and recognition. This chapter tries to solve the problem from graph model and proposes a Self-supervised Feature fusion-based Graph Convolutional Network (SFGCN) model for scene parsing of multi-scale objects.

4.8.1 Experimental Results on PASCAL VOC 2012 Dataset

We train the three variations of the proposed model on the PASCAL VOC 2012 dataset. The SFGCN-SFA architecture incorporated an SFA module into the basic GCN model. The SFGCN variant replaced the standard spatial graph convolution with K-Conv. Additionally, the SFGCN-ALL model added an SDAP module to the latter configuration. The experimental results on the test set of the PASCAL VOC 2012 dataset are presented in Table 4.2, and all of the results are ordered by their mIoU performance from low to high. The proposed method is compared with CRF-based models, for example, DeepLab v2 [7], DeepLab-v3 [2], and other models such as BoxSup [8], DPN [9], PSPNet [3]. The performance of our SFGCN-ALL model achieved a peak accuracy of 79.92%, surpassing both PSPNet and DeepLab-v3. Additionally, the SFGCN-SFA model performed comparably to PSPNet.

Table 4.2 Performance on PASCAL VOC 2012 test set

Methods	mIoU
DeepLab-CRF-LargeFOV-COCO [7]	72.70
MERL-DEEP-GCRF [10]	73.20
POSTECH-DeconvNet-CRF-VOC [11]	74.70
BoxSup [8]	74.80
Context + CRF-RNN [12]	75.20
CentraleSuperBoundaries++ [13]	75.70
DeepLab-CRF-Attention-DT [14]	76.00
DPN [9]	76.80
PSPNet [3]	78.20
DeepLab-CRF+ResNet101 [2]	79.70
SFGCN-SFA	78.18
SFGCN-ALL	**79.92**

4.8.2 Experimental Results on Cityscapes Dataset

For the Cityscapes dataset, this experiment also trained models of four architectures: SFGCN, SFGCN-SDAP, SFGCN-SFA, and SFGCN-ALL. These models were primarily compared with six DCNN-based models, including the basic and extended versions of the SegNet model [4], two versions of DeepLab models [7, 15, 16], as well as other models such as DPN [9], PSPNet [3], PSANet [17], and DenseASPP [18]. Additionally, comparisons were made with a graph convolutional network (GCN) model [19] that utilizes a spatial graph convolution operation.

The mIoU performance of all models is categorized into three groups in descending order based on their types, as shown in Table 4.3. The mIoU performances of the proposed models are 77.20, 77.26, 78.80, and 79.24%, respectively. Among all the comparison models, SFGCN-All ranks third, demonstrating a significant improvement over the GCN model that uses plain graph convolution operations.

Although the performance of SFGCN model is slightly lower than that of PSPNet and PSANet, it is slightly better than that of GCN model, which indicates that the proposed K-Conv operation can improve the accuracy of scene parsing in this dataset. Moreover, after adding SFA to the K-Conv, SFGCN-SFA can improve the accuracy of scene parsing. There is a significant improvement over GCN. In addition, the improvement of SFGCN-SDAP over SFGCN is not obvious, the reason may be that this dataset is collected by fixed equipment, and the consistency of the images is already relatively strong, so the advantage of SDAP module for scene domain adaptation is not obvious.

Table 4.3 Performance on cityscape test set

Methods	mIoU
SegNet basic [4]	57.00
SegNet extended [4]	56.10
DPN [9]	59.10
DeepLab-CRF-LargeFOV [15]	63.10
DeepLab-CRF-LargeFOV-StrongWeak [7]	64.80
PSPNet [3]	78.40
PSANet [17]	80.10
DenseASPP [18]	**80.60**
GCN [19]	76.90
SFGCN	77.20
SFGCN-SDAP	77.26
SFGCN-SFA	78.80
SFGCN-ALL	79.24

4.8.3 Experimental Results on ADE20K Dataset

The ADE20K dataset has a significantly larger number of images and semantic categories compared to the PASCAL VOC 2012 and Cityscapes datasets. The dense image annotations make scene parsing more challenging. On this dataset, two architectures (SFGCN and SFGCN-SDAP) were trained. These models were primarily compared with the SegNet model [4] and its cascade version, Dilated models (DilatedVGG and DilatedResNet) and their cascade versions, as well as other classical models like FCN-8S [1] and PSPNet [3]. Additionally, comparisons were made with newer advanced models such as DMNet [20] and RGNet [21].

The mIoU performance of all models is organized into five groups, ranked from lowest to highest by their type, as shown in Table 4.4. There is considerable room for improvement in the performance of each model on this dataset. The models proposed in this study achieved mIoU performances of 40.20 and 44.30%, respectively. Among these, the SFGCN-SDAP model ranked fourth among all the comparison models, with its performance being slightly lower than that of PSPNet, DMNet, and RGNet.

4.8.4 Experimental Results on COCO-Stuff Dataset

COCO-Stuff is a significantly larger dataset than ADE20K. On this dataset, only the SFGCN-ALL model was trained and compared with DeepLab-CRF [2] and BiSeNetV1 [23]. The performance of all models was arranged in order of mIoU from low to high, as shown in Table 4.5. Compared to BiSeNetV1, the proposed SFGCN-ALL model achieved a higher mIoU of 32.60%, although its pixel accuracy was slightly lower.

Table 4.4 Performance on ADE20K dataset

Methods	mIoU
SegNet [4]	30.51
Cascade-SegNet [22]	32.45
DilatedVGG [12]	37.29
DilatedResNet [12]	37.29
Cascade-DilatedNet [22]	37.79
FCN-8 s [1]	33.44
PSPNet [3]	44.94
DMNet [20]	45.50
RGNet [21]	**45.80**
SFGCN	40.20
SFGCN-SDAP	44.30

Table 4.5 Performance on COCO-stuff dataset

Methods	mIoU	Pix. Acc.
DeepLab-CRF [2]	24.00	58.20
BiSeNetV1 [23]	31.30	**65.50**
SFGCN-ALL	**32.60**	64.60

Table 4.6 Performance on PASCAL-context test set

Methods	mIoU
FCN-8 s [1]	37.80
DeepLab-CRF-LargeFOV [2]	39.60
DeepLab-CRF+ResNet101 [2]	45.70
RefineNet [24]	47.30
PSPNet [3]	47.80
EMANet [25]	53.10
CPNet [26]	53.90
DGCN [19]	53.70
SFGCN-SFA	53.08
SFGCN-SDAP	54.34
SFGCN-ALL	**56.28**

4.9 Ablation Study of SFGCN-Based Scene Parsing Model

In this section, ablation studies of the proposed models are conducted on two datasets, PASCAL-Context and CamVid, to further verify and analyze the efficiency of the three key modules–SFA, K-Conv, and SDAP–on scene parsing performance.

4.9.1 Ablation Study on PASCAL-Context Dataset

The PASCAL-Context dataset is larger than the PASCAL VOC 2012 dataset but smaller than the ADE20K and COCO-Stuff datasets. In this experiment, three network architecture models were trained: SFGCN-SFA, SFGCN-SDAP, and SFGCN-ALL. These models were mainly compared with classic models such as FCN-8S [1], DeepLab [2], RefineNet [24], PSPNet [3], and more recent advanced models like EMANet [25], CPNet [26], and DGCN [19].

The mIoU performance of all models is arranged in three groups from lowest to highest according to their type, as shown in Table 4.6. The performance of the

Table 4.7 Ablation study on PASCAL-context dataset

Models	SFA	K-Conv	SDAP	mIoU
SFGCN	–	–	–	49.20
SFGCN	–	–	✓	50.54
SFGCN	–	✓	–	52.90
SFGCN	–	✓	✓	54.34
SFGCN	✓	–	–	50.10
SFGCN	✓	–	✓	50.80
SFGCN	✓	✓	–	53.08
SFGCN	✓	✓	✓	**56.28**

proposed models are 53.08, 54.34, and 56.28%, respectively. Among all the comparison models, SFGCN-ALL achieved the best scene parsing performance. Furthermore, SFGCN-SFA significantly outperformed SFGCN-SDAP, primarily because the images in this dataset are not specially shot and are often collected through various channels. The acquisition conditions of the original images are greatly influenced by factors such as light, weather, and location, resulting in noticeable differences in image targets within similar scenes. Therefore, compared to the network architecture that only uses the SFA and K-Conv modules, the model with the SDAP module demonstrates higher scene parsing performance.

Additionally, an ablation study was conducted on the PASCAL-Context dataset, with results shown in Table 4.7. This study examined the three main modules: SFA, K-Conv, and SDAP. For the basic GCN, the K-level neighbor of the spatial graph convolution operation was set to $K = 1$, effectively reducing it to a plain convolution operation. In the cases of the SFA and SDAP modules, they were directly removed from the network architecture to assess their individual impacts on performance.

From the experimental results, the model utilizing K-Conv shows the most significant improvement in performance, with mIoU performance generally being better than that of models without K-Conv. Adding either SFA or SDAP modules further enhances model performance, with the SDAP modules providing a greater improvement compared to the SFA modules. Therefore, for the PASCAL-Context dataset, the impact of the SDAP module on the performance of the network model is more pronounced.

4.9.2 Ablation Study on CamVid Dataset

CamVid is a small scene parsing dataset. In this experiment, the SFGCN-SDAP model was trained and compared with the ENet [27], SegNet [4], and BiSeNetV1(R18) [23] models. The mIoU performance of all models is shown in Table 4.8, along with the specific mIoUs for 11 scene semantic categories. The best scene parsing performance in these experiments reached 69.83%, achieved by the SFGCN-ALL model.

Table 4.8 Performance on CamVid dataset

Methods	Building	Tree	Sky	Car	Sign	Road
ENet [27]	74.70	77.80	**95.10**	82.40	51.00	95.10
SegNet [4]	**88.80**	**87.30**	92.40	82.10	20.50	**97.20**
BiSeNetV1(R18) [23]	83.00	75.80	92.00	83.70	46.50	94.60
SFGCN-ALL	82.36	85.92	91.48	**83.74**	**55.48**	89.80
Methods	Pedestrian	Fence	Pole	Sidewalk	Bicyclist	mIoU
ENet [27]	**67.20**	51.70	35.40	**86.70**	34.10	51.30
SegNet [4]	57.10	49.30	27.50	84.40	30.70	55.60
BiSeNetV1(R18) [23]	58.80	**53.60**	31.90	81.40	**54.00**	68.70
SFGCN-ALL	63.46	50.62	**40.36**	85.66	39.28	**69.83**

The specific mIoUs for each model indicate that performance is generally higher for the categories of building, tree, sky, car, road, and sidewalk (mostly above 80%) compared to other categories. This is likely because the images in this dataset are extracted from sequential frames of a continuous street view video. These categories often appear in large pixel areas, and building, sky, road, and sidewalk categories show little variation in the extracted video frames. Additionally, cars and trees frequently appear, leading to a high overlap rate and similarity of samples across training, validation, and test sets, making it easier for the model to learn the pixel features of these categories and achieve better parsing results.

For other semantic categories, especially signs, pedestrians, and poles, the pixel areas are smaller and exhibit greater scale variation due to distance changes during video collection. Consequently, SegNet and BiSeNetV1 models show poor parsing performance for these objects. However, the SFA and K-Conv modules enhance the model's ability to learn local features, resulting in better analytical performance for small-scale objects in the SFGCN and SFGCN-SFA models. Hence, the SFGCN-ALL model shows significant improvement in these categories, which is a major factor contributing to its overall performance.

Additionally, an ablation study was conducted on the CamVid dataset, with results shown in Table 4.9. The ablation setup for each model was the same as that used in the ablation study on the PASCAL-Context dataset. According to the experimental results, the SFA module has the most significant influence on improving model performance. The performance of this model is further enhanced by adding K-Conv or SDAP modules, where the impact of K-Conv module is more obvious on the CamVid dataset.

This chapter introduces the scene parsing method with self-supervised feature fusion-based graph convolution network. Firstly, the hierarchical graph construction is used to construct scene graph feature data in this study, and then the spatial graph convolution operation based on K-level neighbor is used to construct SFGCN model

Table 4.9 Ablation study on CamVid dataset

Models	SFA	K-Conv	SDAP	mIoU
SFGCN	–	–	–	62.60
SFGCN	–	–	✓	63.20
SFGCN	–	✓	–	66.42
SFGCN	–	✓	✓	66.90
SFGCN	✓	–	–	64.50
SFGCN	✓	–	✓	66.86
SFGCN	✓	✓	–	67.10
SFGCN	✓	✓	✓	**69.83**

to achieve global feature learning. Then, to solve the problem that local semantic labels are difficult to parse, a graph attention module based on self-supervised feature fusion is proposed. By combining spectral graph convolution and attention mechanism, self-supervised information is added to model weight training to enhance local feature learning. Finally, in order to preserve the intra-domain consistency of multi-domain scene images, this study proposes a scene domain adaptive model-based graph pooling module to improve the intra-domain inconsistency of scene images caused by their imaging differences. Experimental results on six scene benchmarks show that the proposed method can effectively resolve scene parsing.

References

1. Long, J., Shelhamer, E., Darrell, T.: Fully convolutional networks for semantic segmentation. In: Proceedings of the IEEE Conference on Computer Vision and Pattern Recognition, pp. 3431–3440 (2015)
2. Chen, L., Papandreou, G., Kokkinos, I., Murphy, K., Yuille, A.L.: Deeplab: Semantic image segmentation with deep convolutional nets, atrous convolution, and fully connected crfs. IEEE Transactions on Pattern Analysis and Machine Intelligence 40(4), 834–848 (2018)
3. Zhao, H., Shi, J., Qi, X., Wang, X., Jia, J.: Pyramid scene parsing network. In: Proceedings of the IEEE Conference on Computer Vision and Pattern Recognition, pp. 6230–6239 (2017)
4. Badrinarayanan, V., Kendall, A., Cipolla, R.: Segnet: A deep convolutional encoder-decoder architecture for image segmentation. IEEE Transactions on Pattern Analysis and Machine Intelligence 39(12), 2481–2495 (2017)
5. Li, Z., Li, Y., Tang, L., Zhang, T., Su, J.: Two-person graph convolutional network for skeleton-based human interaction recognition. IEEE Transactions on Circuits and Systems for Video Technology 33(7), 3333–3342 (2023)
6. Cheng, X., He, X., Qiao, M., Li, P., Chang, P., Zhang, T., Guo, X., Wang, J., Tian, Z., Zhou, G.: Multi-view graph convolutional network with spectral component decompose for remote sensing images classification. IEEE Transactions on Circuits and Systems for Video Technology pp. 1–1 (2022)
7. Papandreou, G., Chen, L., Murphy, K.P., Yuille, A.L.: Weakly-and semi-supervised learning of a deep convolutional network for semantic image segmentation. In: International Conference on Computer Vision, pp. 1742–1750 (2015)

8. Dai, J., He, K., Sun, J.: Boxsup: Exploiting bounding boxes to supervise convolutional networks for semantic segmentation. In: International Conference on Computer Vision, pp. 1635–1643 (2015)
9. Liu, Z., Li, X., Luo, P., Loy, C.C., Tang, X.: Semantic image segmentation via deep parsing network. In: International Conference on Computer Vision, pp. 1377–1385 (2015)
10. Vemulapalli, R., Tuzel, O., Liu, M., Chellappa, R.: Gaussian conditional random field network for semantic segmentation. In: Proceedings of the IEEE Conference on Computer Vision and Pattern Recognition, pp. 3224–3233 (2016)
11. Noh, H., Hong, S., Han, B.: Learning deconvolution network for semantic segmentation. In: International Conference on Computer Vision, pp. 1520–1528 (2015)
12. Yu, F., Koltun, V.: Multi-scale context aggregation by dilated convolutions. In: International Conference on Learning Representations (2016)
13. Kokkinos, I.: Pushing the boundaries of boundary detection using deep learning. In: International Conference on Learning Representations (2016)
14. Chen, L., Barron, J.T., Papandreou, G., Murphy, K., Yuille, A.L.: Semantic image segmentation with task-specific edge detection using cnns and a discriminatively trained domain transform. In: Proceedings of the IEEE Conference on Computer Vision and Pattern Recognition, pp. 4545–4554 (2016)
15. Chen, L., Papandreou, G., Kokkinos, I., Murphy, K., Yuille, A.L.: Semantic image segmentation with deep convolutional nets and fully connected crfs. In: International Conference on Learning Representations (2015)
16. Chen, L., Zhu, Y., Papandreou, G., Schroff, F., Adam, H.: Encoder-decoder with atrous separable convolution for semantic image segmentation. In: Proceedings of the European Conference on Computer Vision, vol. 11211, pp. 833–851 (2018)
17. Zhao, H., Zhang, Y., Liu, S., Shi, J., Loy, C.C., Lin, D., Jia, J.: Psanet: Point-wise spatial attention network for scene parsing. In: Proceedings of the European Conference on Computer Vision, vol. 11213, pp. 270–286 (2018)
18. Yang, M., Yu, K., Zhang, C., Li, Z., Yang, K.: Denseaspp for semantic segmentation in street scenes. In: Proceedings of the IEEE Conference on Computer Vision and Pattern Recognition, pp. 3684–3692 (2018)
19. Zhang, L., Li, X., Arnab, A., Yang, K., Tong, Y., Torr, P.H.S.: Dual graph convolutional network for semantic segmentation. In: British Machine Vision Conference, p. 254 (2019)
20. He, J., Deng, Z., Qiao, Y.: Dynamic multi-scale filters for semantic segmentation. In: IEEE International Conference on Computer Vision, pp. 3561–3571 (2019)
21. Yu, C., Liu, Y., Gao, C., Shen, C., Sang, N.: Representative graph neural network. In: Proceedings of the European Conference on Computer Vision, vol. 12352, pp. 379–396 (2020)
22. Zhou, B., Zhao, H., Puig, X., Fidler, S., Barriuso, A., Torralba, A.: Scene parsing through ade20k dataset. In: Proceedings of the IEEE Conference on Computer Vision and Pattern Recognition, pp. 5122–5130 (2017)
23. Yu, C., Wang, J., Peng, C., Gao, C., Yu, G., Sang, N.: Bisenet: bilateral segmentation network for real-time semantic segmentation. In: Proceedings of the European Conference on Computer Vision, vol. 11217, pp. 334–349 (2018)
24. Lin, G., Milan, A., Shen, C., Reid, I.D.: Refinenet: Multi-path refinement networks for high-resolution semantic segmentation. In: Proceedings of the IEEE Conference on Computer Vision and Pattern Recognition, pp. 5168–5177 (2017)
25. Li, X., Zhong, Z., Wu, J., Yang, Y., Lin, Z., Liu, H.: Expectation-maximization attention networks for semantic segmentation. In: International Conference on Computer Vision, pp. 9166–9175 (2019)
26. Yu, C., Wang, J., Gao, C., Yu, G., Shen, C., Sang, N.: Context prior for scene segmentation. In: Proceedings of the IEEE Conference on Computer Vision and Pattern Recognition, pp. 12413–12422 (2020)
27. Paszke, A., Chaurasia, A., Kim, S., Culurciello, E.: Enet: A deep neural network architecture for real-time semantic segmentation. arXiv:1606.02147 (2016)

Chapter 5
Graph Neural Network for Human Parsing

5.1 Introduction

Human parsing, as a fine-grained semantic segmentation task, focuses on accurately identifying various parts of the human body. It plays a crucial role in numerous downstream applications, such as crowd trajectory prediction, action recognition, virtual try-on, and virtual reality. With the advancements in deep Convolutional Neural Networks (CNNs), the performance of human parsing has witnessed significant improvements in recent years. It is worth noting that human bodies exhibit a natural topological physical structure, and different semantic body parts inherently exhibit correlations with one another. Hence, many methods apply graph neural network to this task. The success of deep learning in Euclidean space has motivated researchers to explore non-Euclidean domains, leading to the emergence of Graph Neural Networks (GNNs) [1]. Dai et al. [2] propose an embedding representation for iterative algorithms over graphs, and design a learning method which updates the embeddings and projects them onto the steady-state constraints. Wu et al. [3] propose an innovative adaptive graph representation learning scheme for video person Re-ID, which enables the contextual interactions between relevant regional features. Specifically, they exploit the pose alignment connection and the feature affinity connection to construct an adaptive structure-aware adjacency graph, which models the intrinsic relations between graph nodes. Shi et al. [4] propose a novel Multi-Stream Attention-enhanced Adaptive Graph Convolutional Neural network (MS-AAGCN) for skeleton-based action recognition. The graph topology in this model can be either uniformly or individually learned based on the input data in an end-to-end manner. This data-driven approach increases the flexibility of the model for graph construction and brings more generality to adapt to various data samples. With the general success of the graph neural networks and the natural topological structure of human body, we also attempt to apply GNNs for human parsing.

W. Liu et al., *Graph Neural Network Methods and Applications in Scene Understanding*,
https://doi.org/10.1007/978-981-97-9933-6_5

5.2 Graph Construction for Human Parsing

Human body data is inherently complex and can be organized in hierarchical structures. Understanding and analyzing these structures are crucial for human parsing. In this chapter, we aim to make a more comprehensive understanding for human body and construct a spatial graph and a semantic graph for human parsing. The spatial graph and the semantic graph provide a structured way to model and analyze the human body. Further, by converting human body to graph data, we can utilize the graph neural network to extract more rich features.

5.2.1 Spatial Graph for Human Parsing

Considering the spatial consistency context prior in the human image, we propose to construct a spatial graph for human parsing. Concretely, we define the pixels in the feature map as spatial graph nodes to construct the spatial graph. In this way, we also transform the human parsing task into a graph node classification task. In the spatial graph, the edges between the graph nodes are assigned according to the category of the nodes. Specially, we define the strong-weak attributes for the edges, which can contribute to strengthening the consistency of intra-class features and the discrimination of inter-class features. As shown in Fig. 5.1a, the orange nodes represent the foreground, and the gray nodes represent the background. Based on the definition of the edges, the edges between orange nodes are strong connections, and the gray nodes weakly connect to all other nodes. Here, we define the Fig. 5.1a as coarse-grained spatial graph which only distinguishes the foreground and the background. Further, we construct the fine-grained spatial graph according to the human body parts. In this case, the foreground graph nodes are divided into semantic nodes with categories information. As shown in Fig. 5.1b, the orange nodes are divided into red nodes and blue nodes. The red nodes represent the human body part 1, the blue nodes represent the human body part 2. We can see that the fifth red graph is strongly connected to all red nodes and weakly connected to all gray nodes (background) and blue nodes (human body part 2). Here, the weights of strong-weak connected edges are encoded in the adjacency matrix. As shown in Fig. 5.1, $A_{m,n}$ denotes the weights of edge between the node m and n. We define the $A_{m,n} = 1$ when the node m and n is the same class, else $A_{m,n} = 0$. In this way, we construct the coarse-grained spatial graph and fine-grained spatial graph for human parsing.

5.2.2 Semantic Graph for Human Parsing

Human body presents a natural topological physical structure. Under this circumstance, we aim to construct a semantic graph for human body in the non-Euclidean

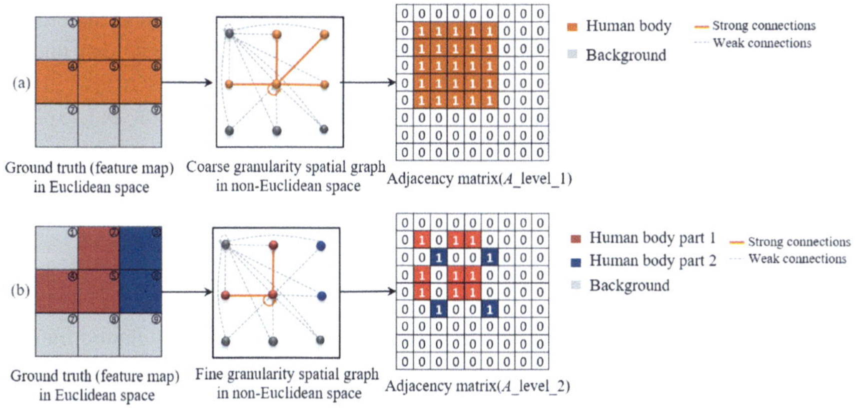

Fig. 5.1 Illustration of different granularity spatial graphs and corresponding adjacency matrixes. In order to show clearly the strong–weak connections between the nodes, we only show the connections of the first and fifth nodes in the graph

space which explicitly formulate the inherent connections. Based on the knowledge prior of human body, we define a semantic graph which the semantic parts and the natural connections are the graph nodes and graph edges, respectively. Specially, this semantic graph is a fixed undirected graph, denoted as $\mathcal{G} = (V, E, A)$, where V and E stand for the set of nodes and edges of semantic graph, and A is the corresponding adjacency matrix. Concretely, $|V| = N$ represents the different semantic parts and N is the number of the labeled categories in the dataset. The edge set E is built based on the inherently correlate of these semantic body parts, and the weights of edges is encoded in the adjacency matrix. As shown in Fig. 5.2, the weight of edge is 1 when the two semantic parts have inherently correlate, and the corresponding

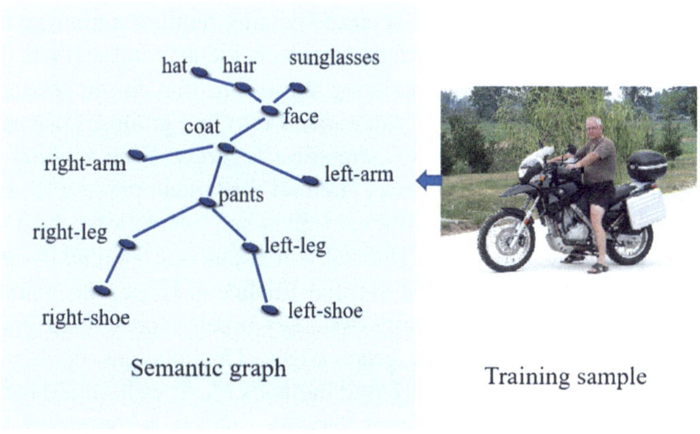

Fig. 5.2 A schematic of fixed graph

$A_{i,j} = 1$, otherwise, $A_{i,j} = 0$, where i and j denotes the graph nodes. Benefiting from the specific topological connection relation of human body, we can understand the human image in the non-Euclidean space. Importantly, the semantic graph enables the model to capture the richer global contextual information and explicit correct semantic structural features.

5.3 Graph Neural Network for Supervised Human Parsing

The human parsing task aims to decompose the human body into semantic human parts and fashion clothes, which has a wide application in computer vision and has attracted a lot of attention in recent years. In this task, each input image is associated with pixel-level annotations specifying the different semantic body parts. As mentioned above, we construct the spatial graph and semantic graph for human parsing. In this chapter, we argue that the hierarchical structure of human body helps to improve the accuracy of human parsing model. Hence, we propose a method which takes full advantage of the hierarchical properties of the human body and the strengths of GNNs.

5.3.1 Graph Convolutional Network Based Human Parsing

The rapid advancement of deep learning has led to substantial progress and success in human parsing. Despite these achievements, human parsing remains a challenging task due to several factors, including the diverse sizes and shapes of human body parts, complex body postures, severe occlusions, and dense crowd scenarios. Notably, human bodies exhibit a natural topological physical structure, and different semantic body parts inherently correlate with each other. These interdependencies make the task even more intricate as accurate parsing requires capturing both local details and global contextual information to precisely identify and segment the human body parts. Given these challenges, existing algorithms that aim to enhance human parsing performance can be broadly categorized into two groups: pose estimation-based and graph-based methods. Pose estimation, regarded as an associated task, is frequently employed as a supplementary method for human parsing [5, 6]. Nevertheless, these algorithms require additional inputs and annotations, such as human poses and key points, leading to heightened computational complexity and potential introduction of further errors. In the non-Euclidean space, the human body's skeleton forms a spatial graph structure, with key points representing graph nodes and natural connections representing graph edges. This inherent topological structure of the human body makes graph-based methods [7, 8] well-suited for this task. Recently, the rise of Graph Convolutional Networks (GCNs) has garnered significant interest in graph-based approaches because they excel in performing effective graph reasoning and offer long-range context information. These methods directly incorporate the topological structure into a fixed graph and subsequently utilize GCNs to propagate information within the graph. Due to the highly structured nature of the human body, numerous methods have been proposed to efficiently model context correlations between body parts using GCN. Gong et al. [7] propose a new universal human parsing agent, named "Graphonomy", which incorporates hierarchical graph transfer learning upon the conventional parsing network to encode the underlying label semantic structures and propagate relevant semantic information. By distilling universal semantic graph representation to each specific task, Graphonomy is able to predict all levels of parsing labels in one system without piling up the complexity. He et al. [8] propose a novel Graph pyramid Mutual Learning (Grapy-ML) method to address the human parsing problem, where the annotations are at different granularities. Starting from the prior knowledge of the human body hierarchical structure, they devise a Graph Pyramid Module (GPM) by stacking three levels of graph structures from coarse granularity to fine granularity subsequently. At each level, GPM utilizes the self-attention mechanism to model the correlations between context nodes. Then, it adopts a top-down mechanism to progressively refine the hierarchical features through all the levels. By making use of the multi-granularity labels, Grapy-ML achieves state-of-the-art performance. Wang et al. [9] argue that modeling the human structure is central for human parsing. They start with analyzing three types of inference processes over the hierarchical structure of human bodies, then formulate the problem as a Compositional Neural Information Fusion (CNIF) framework, which assembles the information from the three inference processes in a conditional manner. Based on CNIF, they further present a Part-Relation-aware Human Parser (PRHP), which precisely describes three kinds of human part relations, and performs iterative reasoning over the human body hierarchy. Enlightened by the verification of the effectiveness of human structure, we also attempt to make full use of the prior knowledge from the human hierarchical structure to facilitate the fine-grained human parsing process.

5.3.2 Graph Neural Network Based Hierarchical Human Parsing

Due to the impressive performance of graph neural networks on structured data and the inherent natural hierarchical structure of human data, we introduce an end-to-end graph neural network based hierarchical human parsing, named Context Prior based Semantic-Spatial Graph Network (CP-SSGNet). The overview of our framework is shown in Fig. 5.3. As depicted in the figure, the algorithm comprises three crucial modules, i.e., semantic constraint module, spatial perceiving module and intra-class attention module, that collaboratively generate hierarchical human parsing results. In general, ResNet-101 is adopted as the backbone of feature extraction. Semantic Constraint Module (SCM) aims to capture contextual information in a global view and mitigate semantic structure errors. To achieve this, we construct a semantic graph

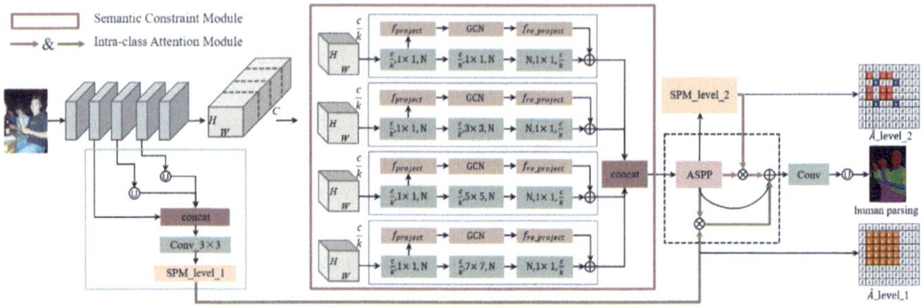

Fig. 5.3 Diagram of the proposed CP-SSGNet. SPM_level_1 and SPM_level_2 represent the SPM composed of the coarse granularity spatial graph and the fine granularity spatial graph respectively. $\hat{A}\_level\_1$ is the predicted adjacency matrix of the coarse granularity spatial graph. $\hat{A}\_level\_2$ is the predicted adjacency matrix of fine granularity spatial graph

from the high-level features which explicitly represents the inherent topological connection relationship with each node corresponding to a specific human body part. The construction of the semantic graph is elaborated in Sect. 5.2.2. Additionally, drawing inspiration from the split-transform-merge strategy [10, 11] and group convolution [12], we partition the input channels into K groups and incorporate the semantic graph with varying sizes of convolution kernels within each group. By doing so, the SCM generates more comprehensive global contextual features while maintaining manageable computational costs.

Then, we introduce the Spatial Perceiving Module (SPM), which aims to enhance local features by incorporating spatial consistency context prior through a spatial graph. As detailed in the Sect. 5.2.1, each pixel in a feature map is treated as a graph node, forming a spatial graph. Notably, we define strong-weak connections in the spatial graph to encode spatial consistency context, and leverage Graph Convolutional Networks (GCNs) to reason about the spatial graph and obtain an improved spatial graph model. By combining the features of a vertex with those of its neighboring nodes, the advanced spatial graph model learns enriched local features, effectively reducing local consistency errors. As a result, SPM plays a vital role in improving the accuracy and coherence of the human parsing results. Importantly, to generate hierarchical human parsing results, we create two types of spatial graph with different granularity levels. The coarse granularity spatial graph is generated using low-level features, while the fine granularity spatial graph is generated using high-level features. This approach allows us to capture different levels of detail and context of the human body. And then we can achieve a hierarchical understanding of the human body and obtain more comprehensive human parsing results. Lastly, to further augment the intra-class features of human parsing, we introduce the Intra-class Attention Module (IAM), which is supervised directly by the output of SPM. Unlike conventional attention modules that learn implicit attention weights through the network, IAM utilizes the adjacency matrix of the spatial graph as attention weights. By leveraging this explicit intra-class context, IAM aggregates specific intra-class features. The strong-weak connections encoded in the adjacency matrix capture the essential relationships between body parts, enabling IAM to focus on relevant information and enhance the representation of intra-class features. This design enables IAM to refine and highlight important details of the human body, thereby improving the accuracy and precision of the parsing results. Once we obtain three predictions from inference networks, and the hierarchical human parsing training loss can be formulated as:

$$\mathcal{L} = \alpha \mathcal{L}_{coarse} + \beta \mathcal{L}_{fine} + \lambda \mathcal{L}_{parsing}, \tag{5.1}$$

where α, β and λ are hyper-parameters to balance these three losses.

5.3.3 Framework and Procedure

Since the human body exhibits a strong hierarchical structure, some researchers propose hierarchical human semantic parsing methods and focus on analyzing the effectiveness of the highly structured hierarchy of the human body in boosting performance. Based on the hierarchical semantic structural prior knowledge of the human body, we construct both the coarse-grained spatial graph and fine-grained spatial graph for human parsing, as shown in Fig. 5.1. Here, we aim to design a universal human parsing model that can parse the human body into different label granularity.

Semantic Constraint Module. Unlike other objects, the human body has a specific topological connection relationship and semantic structure context prior. Capturing and comprehending contextual information in a global view can effectively reduce semantic structure errors, especially for challenging cases like limb absence or complex poses. Some previous methods implicitly attempted to extract structural information by deepening convolutional layers, but this approach could not adequately capture correct and richer global contextual information. To fully leverage the power of the human body's semantic structure context prior, we introduce the Semantic Constraint Module (SCM). SCM constructs a semantic graph from the high-level features, explicitly formulating the inherent topological connection relationship. Drawing inspiration from ASPP [13] and InceptionNets [14, 15], which aggregate multi-scale information with different kernel sizes, we adopt the split-transform-merge strategy [12, 13] and employ multiple parallel filters with varying sizes. This allows us to generate more comprehensive global contextual features while maintaining acceptable computational costs. In Fig. 5.3, the high-level feature maps are split into four groups, and the parallel kernel sizes are 1×1, 3×3, 5×5, and 7×7. Next, we provide further details on the process of SCM to illustrate its role in capturing global contextual information for improved human parsing.

The human body exhibits a distinct semantic structure context prior, where different semantic parts possess fixed connection relationships. For example, the head, arms, and legs are connected to the torso. To capture this semantic structure context, we define an undirected graph as $G = (V, E, A)$, where V is the graph nodes with $|V| = N$, N is the number of classes. E is the edge set representing the fixed connection relationships, and A is the adjacency matrix. To start the process, we split the output feature maps of ResNet-101 $X \in \mathbb{R}^{H \times W \times C}$ into K groups, as follows:

$$X \rightarrow X_g \in R^{H \times W \times \frac{C}{K}}, g \in [1, 2, \ldots, K], \tag{5.2}$$

thus, $X_g \in R^{H \times W \times \frac{C}{K}}$ is the input of the SCM, where H, W and $\frac{C}{K}$ are height, width and channel number of the feature maps, as shown in Fig. 5.3. Following a 1×1 convolutional layer, we design a projection function to produce semantic graph data $S_g^{se} \in R^{N \times P}$ as follows:

$$X_g^p = X_g \circledast f_\theta, \tag{5.3}$$

$$S_g^{se} = f_{projection}(X_g^p, \omega), \tag{5.4}$$

where $\circledast$ denotes the convolution operation, $X_g^p \in R^{H \times W \times N}$ is the output of the 1×1 convolutional layer, $\omega \in R^{Z \times P}$ is the trainable transformation matrix to convert X_g^p into S_g^{se}, and $Z = H \times W$. After the initial splitting of feature maps into K groups, we proceed to extract enhanced global representation through the use of multiple graph convolutional layers. These graph convolutional layers can be denoted as follows:

$$H_{l+1} = \sigma(\widetilde{D}^{-\frac{1}{2}} \widetilde{A} \widetilde{D}^{-\frac{1}{2}} H_l W_l), \tag{5.5}$$

where, $\widetilde{A} = A + I$ is the adjacency matrix of the undirected graph G with added self-connections. I is the identity matrix, $\widetilde{D}_{ii} = \sum_j (\widetilde{A}_{i,j})$ is a diagonal matrix of node degrees, $W_l \in R^{P_l \times P_{l+1}}$ is a learnable parameter matrix, $\sigma(\cdot)$ denotes a nonlinear activation function. After that, we again apply another projection function $f_{re\_project}(\cdot)$ to re-project the semantic graph nodes to feature maps X_g^{re}, followed by a summation function, as follows:

$$X_g^{re} = f_{re\_projection}(GCN(S_g^{se}), \omega^{re}), \tag{5.6}$$

$$X_g{}' = X_g^{re} \oplus conv(X_g^p), \tag{5.7}$$

where, the operation $conv(\cdot)$ represents multi-layer convolution. We apply different convolutional kernel sizes in the group convolution to extract multi-scale features, as illustrated in Fig. 5.3. This approach allows the SCM to effectively capture semantic structure context information in a global view, enhancing its ability to understand the intricate relationships between different semantic parts in the human body. As a result of this process, the outputs of SCM gain a stronger capacity to represent and utilize the semantic structure context information, contributing to improved human body analysis results.

Group convolution has been adopted and demonstrated to be efficient in subsequent research [12]. In this chapter, we combine the split-transform-merge strategy and multiple parallel filters with different sizes to generate more comprehensive global contextual features while ensuring computational efficiency. Finally, all the outputs of each group are aggregated by concatenating them together. This step can be expressed as follows:

$$X' = \left[X_1', X_2', \ldots, X_K'\right]. \tag{5.8}$$

By doing so, we obtain a holistic and enriched representation of the global contextual information from the multiple group convolutions, allowing for effective human parsing within acceptable computational costs.

Spatial Perceiving Module. As shown in Fig. 5.3, we introduce a Spatial Perceiving Module (SPM), which includes SPM_level_1 and SPM_level_2, to capture local features by encoding the spatial consistency context prior. As mentioned before, we have defined two kinds of granularity spatial graph, i.e., coarse-grained spatial graph and fine-grained spatial graph. Specially, coarse-grained spatial graph only segments the human body and background. In this case, the coarse-grained spatial graph can provide accurate edge features. The fine-grained spatial graph, which labels the human body into different body parts and fashion clothes, can exploit comprehensive and accurate local consistency features. Concretely, we define an undirected graph as $g = (V, E, A)$, where V and E stand for the set of nodes and edges of semantic graph, and A is the corresponding adjacency matrix. Concretely, $|V| = Z$ represents the graph nodes and Z is the number of the pixels in feature maps. The architecture of graph neural network based hierarchical human parsing is illustrated in Fig. 5.3. Graph neural network based hierarchical human parsing model learns highly precise and detailed boundaries features with SPM_level_1 (the coarse-grained spatial graph), and rich semantic features with SPM_level_2 (the fine-grained spatial graph). Previous methods extract features with the classical ResNet-101 which has been verified the effectiveness on multiple visual tasks. On the basis of this verification, graph neural network based hierarchical human parsing model also applies the ResNet-101 as the backbone to extract features for the hierarchical human parsing. As we can see in Fig. 5.3, there are two branches for the coarse-grained human parsing and fine-grained human parsing, respectively. For the coarse-grained human parsing, SPM_level_1 utilizes the low-level visual features (Res2, Res3, Res4) from ResNet-101 to construct the coarse-grained spatial graph. We firstly upsample the Res3 and Res4 into the size of Res2, and concatenate these feature maps to generate the fusion features, as follows:

$$F_{fusion} = Conv(concat(\text{Res2}, up(\text{Res3}), up(\text{Res4}))). \tag{5.9}$$

Further, the fusion features F_{fusion} are transformed to coarse-grained spatial graph $G_{coarse} \in \mathbb{R}^{Z \times D}$,

$$G_{coarse} = F^{re}_{fusion} W_{coarse}, \tag{5.10}$$

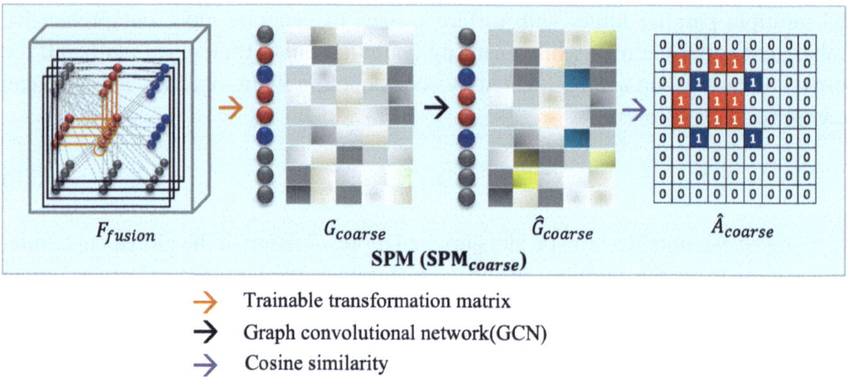

Fig. 5.4 Illustration of Spatial Perceiving Module. G_{coarse} denotes the reshaped spatial graph. For visual clarity, we only show the fine granularity spatial graph

where F_{fusion} is reshaped into $F^{re}_{fusion} \in \mathbb{R}^{Z \times C}$, $Z = H \times W$, and $W_{coarse} \in \mathbb{R}^{C \times D}$ is the trainable transformation matrix. Since previous researchers have revealed the powerful feature extraction ability of GCN for graph data, two layers of GCN are deployed for graph reasoning to obtain an advanced coarse-grained spatial graph $\hat{G}_{coarse}$, as shown in Fig. 5.4. Here, the GCN is same as the $*$, and we utilize the cosine similarity to compute the cosine similarity between the coarse-grained spatial graph nodes, as follows:

$$A_{m,n} = \frac{\hat{G}_{coarse,m} \cdot \hat{G}_{coarse,n}}{\left\|\hat{G}_{coarse,m}\right\| \left\|\hat{G}_{coarse,n}\right\|}, \tag{5.11}$$

where $\hat{G}_{coarse,m}$ and $\hat{G}_{coarse,n}$ denote the tensor of the coarse-grained spatial graph nodes, respectively. $\|\cdot\|$ computes the value the tensor, and $\cdot$ is the element-wise product. By computing the cosine similarity of all coarse-grained spatial graph nodes, we can generate the predicted coarse-grained adjacency matrix $\hat{A}_{coarse} \in \mathbb{R}^{Z \times Z}$. Specially, this predicted coarse-grained adjacency matrix $\hat{A}_{coarse} \in \mathbb{R}^{Z \times Z}$ also participate in the final loss. Concretely, we design a spatial graph perceiving based on the binary cross entropy loss:

$$\mathcal{L}_{A_{coarse}}(\mathcal{L}_{A_{fine}}) = -\frac{1}{Z^2}\sum_{n=1}^{Z^2}(a_n log\hat{a}_n + (1 - a_n)log(1 - \hat{a}_n)), \tag{5.12}$$

where $\{a_n \in A_{coarse}, n \in [1, Z^2]\}$ is the adjacency matrix generated by the ground truth, and $\left\{\hat{a}_n \in \hat{A}_{coarse}, n \in [1, Z^2]\right\}$ presents the predicted adjacency matrix.

Intra-class Attention Module. After the spatial perceiving module, we further design an Intra-class Attention Module (IAM) to generate the final hierarchical human parsing results. In human parsing, correctly learning and understanding the intra-class consistency and inter-class distinction is essential for improving the accuracy of segmentation results. To achieve this, one popular and important measure is to learn implicit attention weights which will aggregate intra-class features and reinforce inter-class distinction. To efficiently capture the powerful intra-class features and discriminate inter-class representations for optimizing the human parsing results, IAM directly sets the predicted adjacency matrix of the spatial graph as the attention weights. As defined in function 5.4, this attention weights can be supervised by the adjacency matrix of ground truth. Hence, instead of implicitly learning attention weights, IAM generates attention weights in an explicitly supervised manner, which is considered that this way can more efficiently and accurately aggregate intra-class features and strengthen inter-class features. The structure of IAM is illustrated in Fig. 5.3. Firstly, the traditional ASPP is utilized to further produce multi-range spatial information, and then the output of the ASPP is reshaped to $F \in \mathbb{R}^{Z \times N}$. After this, we perform two times attention mechanism according to the predicted coarse-grained adjacency matrix $\hat{A}_{coarse} \in \mathbb{R}^{Z \times Z}$ and fine-grained adjacency matrix $\hat{A}_{fine} \in \mathbb{R}^{Z \times Z}$, respectively.

$$F_{coarse} = reshaped(F^T \hat{A}_{coarse}), \tag{5.13}$$

$$F_{fine} = reshaped(F^T \hat{A}_{fine}). \tag{5.14}$$

For fine-grained human parsing, the model is required to segment the body parts with highly detailed boundaries and different semantic region as accurately as possible. In this case, we generate the rich and accurate features for fine-grained human parsing by integrating features from three branches via an element-wise summation, as follows:

$$F'_{fine} = sum(F_{coarse}, F_{fine}, F). \tag{5.15}$$

For the coarse-grained human parsing, considering that there are only two categories, foreground (human body) and background, which does not need to classify different semantic region, we directly employ the generated F_{coarse} as the final features to predict the coarse-grained human parsing. Finally, designing a decoder module by deploying a convolution layer and an upsampling layer, we produce the coarse-grained human parsing results and fine-grained human parsing results, as follows:

$$\hat{y}_{coarse} = upsample(conv(F_{coarse})), \tag{5.16}$$

$$\hat{y}_{fine} = upsample(conv(F'_{fine})), \tag{5.17}$$

where $\hat{y}_{coarse}$ and $\hat{y}_{fine}$ are the predicted coarse-grained human parsing results and fine-grained human parsing results, respectively. In the training process, the hierarchical human parsing loss is defined as:

$$\mathcal{L}_{parsing} = \alpha \mathcal{L}_{coarse} + \beta \mathcal{L}_{fine}, \tag{5.18}$$

$$\mathcal{L}_{coarse} = \frac{1}{pix} \sum_{pix} \sum_{N_c} -y_{coarse} log \hat{y}_{coarse}, \tag{5.19}$$

$$\mathcal{L}_{fine} = \frac{1}{pix} \sum_{pix} \sum_{N_f} -y_{fine} log \hat{y}_{fine}, \tag{5.20}$$

where y_{fine} and y_{coarse} are the original ground truth with different granularity, pix denotes the number of pixels in images, and N_c and N_f is the number of classes.

5.3.4 Experimental Results and Discussions

Results on PASCAL-Person-Part. The PASCAL-Person-Part dataset contains 3,535 annotated images distributed to 1,717 for training and 1,818 for testing, among which only 7 categories are labeled including 6 human body parts (head, torso, upper/lower arms, upper/lower legs) and the background. This dataset is challenging due to the small-scale dataset. In Table 5.1, the designed model is evaluated with several state-of-the-art methods. As we can see, the designed model achieves the best performance with a 72.19% mIoU. Some previous methods are able to generate a high performance with over 70% mIoU, such as Graphonomy and GPM. Specially, these methods train the model across different datasets to improve model performance by leveraging the heterogeneous multi-granularity annotations. Different from these methods, the designed model aim to make full use of the multi-level structure context prior knowledge of human body to boost the performance, and the quantitative results have demonstrated the validity of the proposed model. In addition, we can observe that the designed model are able to generate the best segmentation results on Torso, upper-leg and lower-leg, which are marked in bold in Table 5.1. Although the designed model performs slightly worse on the other semantic categories, it learns rich semantic features to produce the highest mIoU. In addition, qualitative comparisons of the proposed method and representative methods on PASCAL-Person-Part dataset are conducted by showing some human parsing results in Fig. 5.5. We can observe that the human parsing performance is affected by the challenging poses and complex backgrounds.

Results on LIP. LIP dataset is currently a large-scale benchmark dataset which contains 50,462 images, including 30,462/10,000/10,000 images for training/verification/testing, respectively. It defines elaborated pixel-level annotations of 19 semantic human part labels, including 13 clothes categories and 6 body parts. Furthermore, all of the images in LIP are collected from real-world scenarios that contain heavy occlusions, various appearances, and complex poses. To verify the performance of the designed model, we compare our framework with other deep-learning-

Table 5.1 Comparison of human parsing performance with several state-of-the-arts methods on PASCAL-person-part dataset

Methods	Head	Torso	u-arm	l-arm	u-leg	l-leg	bkg	mIOU
Attention (2016) [16]	81.47	59.06	44.15	42.50	38.28	35.62	93.65	56.39
Part FCN (2017) [17]	85.50	67.87	54.72	54.30	48.25	44.76	95.32	64.39
MMAN (2018) [18]	82.58	62.83	48.49	47.37	42.80	40.40	94.92	59.91
WSHP (2018) [19]	87.15	72.28	57.07	56.21	52.43	50.36	**97.72**	67.60
PGN (2018) [20]	**90.89**	75.12	55.83	64.61	55.42	41.57	95.33	68.40
DeepLabV3+ (2018) [21]	–	–	–	–	–	–	–	68.60
JPPNet (2019) [5]	83.26	62.40	47.80	45.58	42.32	39.48	94.68	59.36
Learning (2019) [22]	88.02	72.91	64.31	63.52	55.61	54.96	96.02	70.76
Graphonomy (2019) [8]	–	–	–	–	–	–	–	71.14
A-CE2P (2019) [23]	87.41	73.80	**64.98**	**64.70**	57.43	55.62	96.26	71.46
GPM (2020) [7]	–	–	–	–	–	–	–	71.65
Ours	88.93	**75.81**	62.87	62.44	**60.37**	**58.30**	96.62	**72.19**

based methods, as shown in Table 5.2. As we can see, the designed model yields a high performance with 54.90% mIoU, which is slightly lower than the best model CorrPM with 55.33% mIoU. The underlying reason is that CorrPM combined three sub-tasks (edge, pose and parsing) into a unified network. The CorrPM needs additional key point annotations of the human body while obtaining high performance. The proposed model attempt to optimize the framework with the multi-level structure context prior information of the human body, and the quantitative results verify that the proposed model achieves a competitive result compared to CorrPM with acceptable computational costs. CorrPM† deleting the heterogeneous non-local module performs worse than the proposed model with 0.47% mIoU. Benefitting from leveraging the feature resolution, global context information, and edge details, CE2P undoubtedly obtains a high mIoU with 53.10%. Whereas both DeepLab (VGG-16) and Attention performed slightly worse. This is a reasonable results as the human parsing requires advanced semantic features to obtain accurate segmentation accuracy. Compared with DeepLab (VGG-16), DeepLab (ResNet-101) can learn advanced semantic representation by deepening the convolutional layers. Based on the above analysis, our algorithm also adopts ResNet-101 as the backbone, and achieves the best experimental results.

The value of P **and** D **in Eqs.** (5.3) **and** (5.9)**.** In Eqs. (3) and (9), the hyper-parameters P and D represent the dimensions of the graph nodes. In this chapter, we maintain uniformity in the values of hyper-parameters P and D. The proposed model, including both the SCM and SPM, incorporates two layers of GCNs, denoted

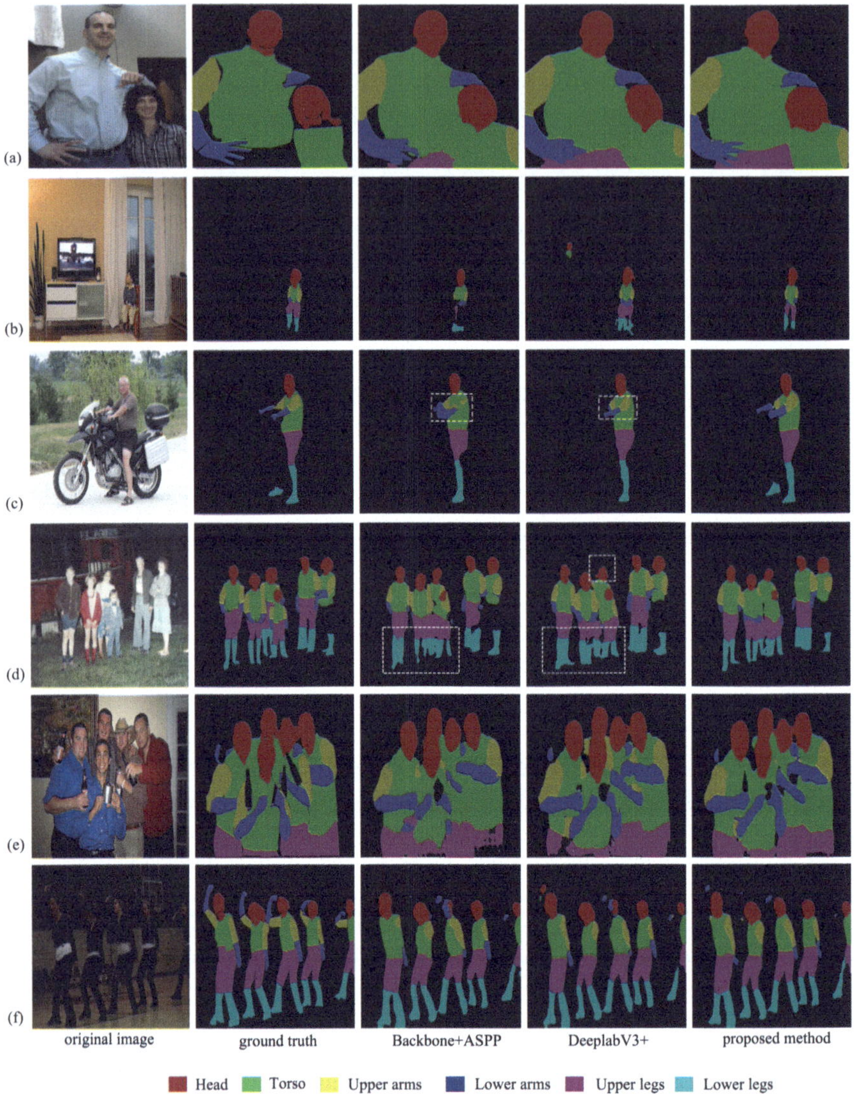

Fig. 5.5 Visual results on the PASCAL-Person-Part test set. The five columns in the image are the original image, ground truth, Backbone + ASPP, DeeplabV3+, proposed method, respectively

as GCN1 and GCN2. During the graph reasoning process, we maintain a fixed number of graph nodes and only vary the dimension of the graph nodes. We empirically set the dimension of graph nodes as 32, 64, 128, or 256. The results obtained on the LIP validation set are summarized in Table 5.3. Here, $P = D$ denotes the dimension of the input graph nodes, while GCN1_output and GCN2_output signify the dimensions of the output graph nodes in GCN1 and GCN2, respectively.

Table 5.2 Comparison of pixel accuracy, mean accuracy and mIoU on LIP validation set [15]

Method	Publish	Pixel accuracy	Mean accuracy	mIoU
Attention [16]	2016CVPR	83.43	54.39	42.92
DeepLab (VGG-16) [13]	2018T-PAMI	82.66	51.64	41.64
DeepLab (ResNet-101) [13]	2018T-PAMI	84.09	55.62	44.80
MMANs [18]	2018ECCV	–	–	46.81
MuLA [24]	2018ECCV	88.5	60.5	49.3
SS-JPPNet [5]	2019T-PAMI	84.36	54.94	44.73
JPPNet [5]	2019T-PAMI	86.39	62.32	51.37
CE2P [25]	2019AAAI	87.37	63.20	53.10
CorrPM† [26]	2020CVPR	87.36	66.37	54.43
CorrPM [26]	2020CVPR	**87.68**	**67.21**	**55.33**
Ours	–	87.67	66.52	54.90

Table 5.3 Results of our model with various values of P and D on LIP validation set [15]

P/D	GCN1_output	GCN2_output	Pixel accuracy	Mean accuracy	mIoU
128	256	64	85.02	65.07	52.21
128	128	64	85.89	65.16	52.31
128	64	64	85.90	65.27	52.77
64	64	64	87.44	66.25	54.20
32	32	32	86.14	65.90	53.02
64	64	32	87.51	66.18	54.04
64	32	64	87.54	66.34	54.56
32	**64**	**64**	**87.67**	**66.52**	**54.90**

The group number K. In this chapter, we implement multiple branches, each employing different kernel sizes, to aggregate a broader range of multi-scale information. As previously mentioned, the feature map is divided into K groups to facilitate this approach. To study the effect of hyper-parameter K on the final experimental results, we conducted experiments on the various values of hyper-parameters K, as shown in Table 5.4. It is a well-established practice to employ kernel sizes of 3×3, 5×5, and 7×7 in previous experiments. Motivated by this, we empirically select the kernel size and set the number of groups to 1, 2, and 4. The results presented in Table 5.4 affirm the advantageous impact of utilizing multiple kernels.

Visualization of the predicted adjacency matrix. To facilitate an intuitive understanding of the adjacency matrix (attention weights), visualizations of randomly selected predicted adjacency matrices and corresponding ideal adjacency matrices from the PASCAL-Person-Part test set are depicted in Figs. 5.6 and 5.7. As previously noted, the size of the ground truth map is 321×321. To construct the adja-

Table 5.4 Results of our model with various values of K on LIP validation set [15]

K	Kernel size	Pixel accuracy	Mean accuracy	mIoU
1	3×3	87.28	66.03	53.54
2	$3 \times 3, 5 \times 5$	87.29	66.24	53.78
4	$\mathbf{1 \times 1, 3 \times 3, 5 \times 5, 7 \times 7}$	**87.67**	**66.52**	**54.90**

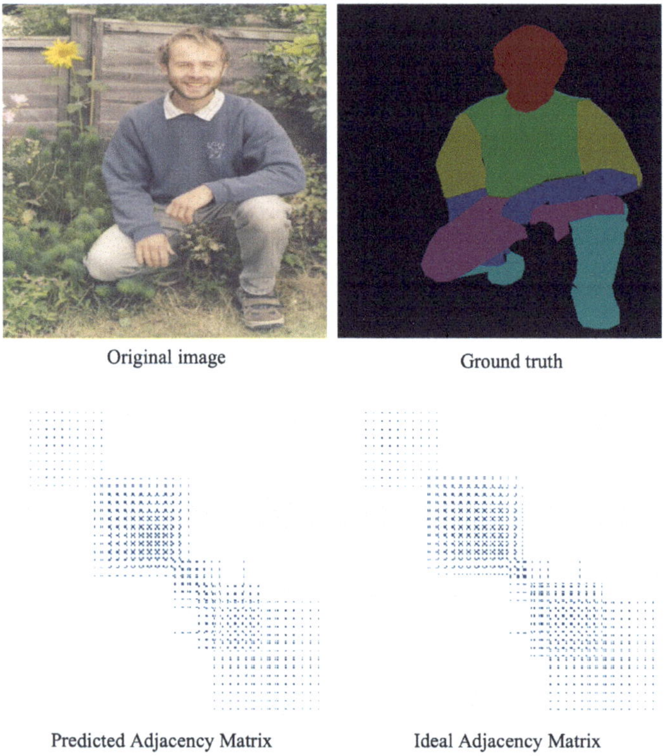

Fig. 5.6 Visual results on the PASCAL-Person-Part test set. The five columns in the image are the original image, ground truth, Backbone + ASPP, DeeplabV3+, proposed method, respectively

cency matrix, we sub-sample the ground truth map to a size of 41×41, resulting in an ideal adjacency matrix of dimensions 1681×1681. In the IAM, we set the attention weights as the adjacency matrixes. Utilizing the ideal adjacency matrix, our model is able to train attention weights under supervision. In comparison to the learned implicit attention weights, our attention weights encode more explicit structure information.

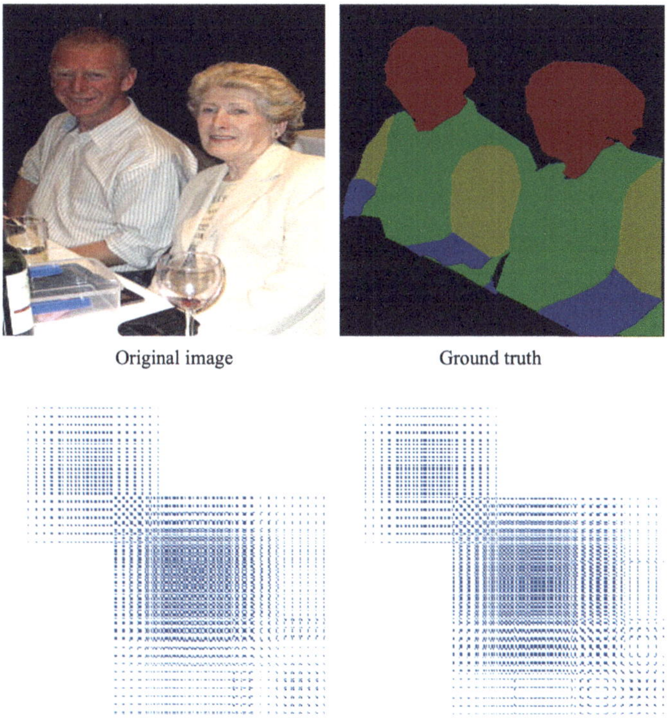

Fig. 5.7 Visual results on the PASCAL-Person-Part test set. The five columns in the image are the original image, ground truth, Backbone + ASPP, DeeplabV3+, proposed method, respectively

Table 5.5 Ablation experiments in PASCAL-person-part test

Backbone+ASPP	SCM	SPM	IAM	Pixel accuracy	Mean accuracy	mIoU
√	–	–	–	93.67	73.49	64.94
√	√	–	–	94.87	78.30	70.89
√	–	√	–	93.82	75.16	67.41
√	–	√	√	95.06	79.31	71.27
√	√	√	√	**95.37**	**81.22**	**72.19**

Ablation study for semantic constraint module. As shown in Table 5.5, there is a performance gap of 7.25% between the proposed method and the Backbone + ASPP. When only adopting the Semantic Constraint Module (SCM), the gap reduces to 5.95%. This demonstrates that the SCM contributes the most to the accuracy improvement among the three modules. Table 5.6 further reveals the impact of SCM. With adding the SCM, the accuracy of SPM_level_1 + IAM improves from 68.35 to 71.80%, and that of SPM_level_2 + IAM increases from 69.84 to 71.64%. The

Table 5.6 Ablation experiments for SPM_level_1 and SPM_level_2 in PASCAL-person-part test

Backbone+ ASPP	SCM	SPM_level_1	SPM_level_2	IAM	Pixel accuracy	Mean accuracy	mIoU
√	–	–	√	√	94.80	77.39	69.84
√	–	√	–	√	94.14	76.94	68.35
√	√	–	√	√	95.01	79.89	71.64
√	√	√	–	√	95.12	80.04	71.80
√	√	√	√	√	**95.37**	**81.22**	**72.19**

significant improvements are attributed to the rich semantic structure features captured by SCM, enabling effective handling of semantic structure errors, particularly in challenging scenarios involving complex postures or missing limbs. The SCM proves to be particularly beneficial in handling hard samples, leading to overall enhanced human body analysis results. There are severe category imbalances and a substantial number of human pictures with limb absence, and the SCM plays a crucial role in addressing these challenges. SCM constructs a semantic graph from the high-level feature map, and makes full use of the inherent structural context prior of the human body. By explicitly modeling the topological connection relationships between different semantic parts, SCM enables the model to capture and leverage the rich structural information present in the human body. Consequently, SCM helps eliminate common structural errors and enhances the overall performance of the model. This is particularly beneficial in scenarios with imbalanced categories and cases of limb absence, where SCM effectively leverages the structural context to improve the accuracy and reliability of the analysis. In Fig. 5.8, we present visual results from the ablation experiments on the LIP dataset. The LIP dataset contains 19 semantic human parts and notably distinguishes between left and right hands, left and right legs, and other similar body parts. Such similar appearances can pose challenges and lead to structural errors in the models, such as incorrect discrimination between left and right parts. However, the SCM effectively addresses this issue by explicitly capturing and leveraging the inherent topological connection relationships between different semantic parts. By doing so, SCM ensures that the model can better understand and differentiate between left and right parts, thus significantly reducing structural errors and improving the accuracy of the human body analysis. The visual results demonstrate the effectiveness of SCM in handling such complex distinctions and enhancing the overall performance of the model on the challenging LIP dataset. As illustrated in Fig. 5.8a, for challenging back view samples, the method comprising Backbone + ASPP + SCM accurately parses the left arm and right arm, despite failing to segment the feet. Similarly, in Fig. 5.8b and c, the method correctly labels the left foot and right foot, although it may not recognize the arms. Indeed, the visual results presented in Fig. 5.8 provide strong evidence and verification of the effectiveness of the SCM. However, despite the excellent global feature representation ability of the SCM, there are some structural errors in Fig. 5.8d column 4. If there are confused

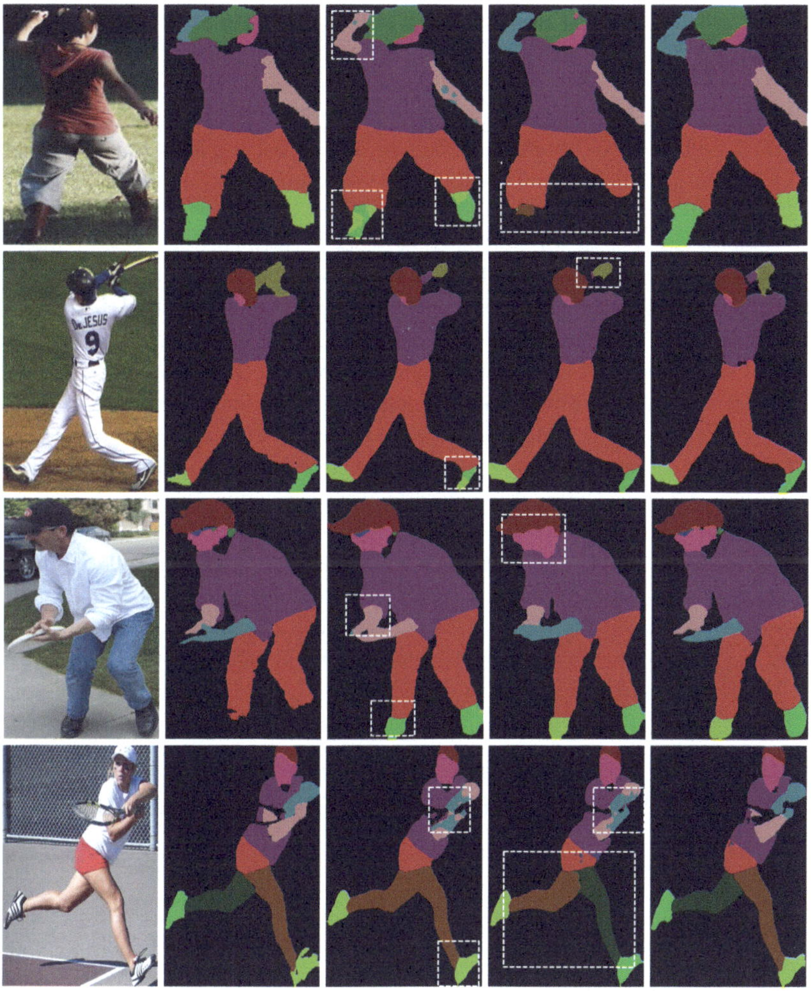

Fig. 5.8 Visual results on LIP validation set. From left to right are the original image, ground truth, Backbone + ASPP + SPM + IAM, Backbone + ASPP + SCM, and the proposed method, respectively

mirror categories, Backbone + ASPP + SCM cannot accurately identify the left leg and right leg.

Ablation study for Spatial Perceiving Module and Intra-class Attention Module. Similar to the Laplacian smoothing, the SPM technique combines the features of a vertex with those of its adjacent nodes. Consequently, this process facilitates the clustering of similar features and effectively diminishes local consistency errors, including noise points commonly encountered in segmentation maps. Take advantage of these means, SPM yields an mIOU of 67.41% and increases 2.47% compared to the Backbone+ASPP. Nevertheless, SPM proves to be inferior to SCM in terms of enhancing network performance, as demonstrated by an mIOU of 70.89% for SCM in Table 5.5. As we have elaborated before, SCM exhibits the ability to extract comprehensive and meaningful semantic structural information from a global perspective. The PASCAL-Person-Part dataset and LIP dataset encompass numerous challenging samples, such as instances with heavy occlusions, missing limbs, and intricate poses. In this context, an abundance of accurate and intricate semantic structural information can significantly enhance network performance. However, when integrating the SPM and IAM into the Backbone+ASPP architecture, the model yields a noteworthy enhancement, reaching 71.27%, as illustrated in Table 5.5. The IAM employs a spatial graph with strong-weak connections to aggregate intra-class features effectively, enabling a clearer distinction of inter-class features. Through this approach, the IAM implicitly captures and extracts semantic structure information at the spatial structure level. Furthermore, the majority of previous methods have relied on learned implicit attention weights to aggregate intra-class features, leading to limited intra-class consistency and inter-class distinction. In contrast, this chapter introduces a novel approach where the outputs of SPM serve as direct supervision for the attention weights in IAM. As a result, SPM + IAM is capable of capturing diverse intra-class features and effectively discriminating inter-class representations. Consequently, in comparison to SCM, which solely extracts semantic structure features, the fusion of local consistency features and semantic structure features in SPM + IAM enables it to achieve higher precision.

5.4 Graph Neural Network for Weakly Supervised Human Parsing

Based on the natural topology of human body, graph neural network plays a great role in improving the accuracy of human parsing model. Further, we try to explore the possibility of graph neural network model in weakly supervised human parsing. For weakly supervised human parsing, the weakly supervised information can be image-level information, box-level information, scribble-level information and point-level information. With the box-level labels and point-level labels, we propose a two-stage learning framework, by combining generation pseudo labels and graph neural network based weakly supervised human parsing.

5.4.1 Generation Pseudo Labels for Human Parsing

In this chapter, we aim to generate the pixel-level pseudo labels for human parsing only with some weak supervision information. There are some methods to alleviate the pixel-level annotation burden by introducing some pseudo labels generation algo-

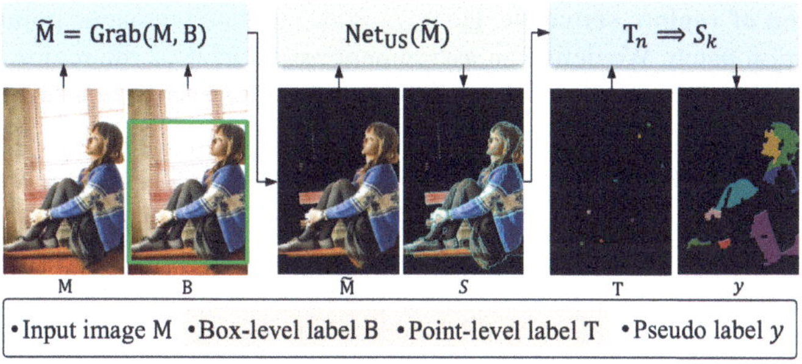

Fig. 5.9 An example of generated pixel-level pseudo label with region-semantic based unsupervised segmentation strategy

rithms, which can be broadly classified into four formats i.e. image-level supervision, box-level supervision, scribble-level supervision and point-level supervision. Human parsing is a fine-grained semantic segmentation and needs to disentangle different body parts as well as various fashion clothes from the holistic body. Additionally, multiple small objects further aggravate the difficulty of generating pseudo labels. Under the above analysis, we attempt to apply both box-level labels and point-level labels as the weak supervision. Concretely, we design a region-semantic based unsupervised segmentation strategy to generate the pixel-level pseudo labels, as shown in Fig. 5.9. Initially, the input images M are preprocessed by the Grab-cut with the box-level labels B. Grab-cut is a high-efficiency image segmentation algorithm, which can segment the input images into the foreground (the holistic human body) and background, as follows:

$$\tilde{M}_i = Grab(M, B) = \begin{cases} M_i, & humanbody \\ 0, & background \end{cases}, \tag{5.21}$$

where $\tilde{M}_i$ is the value of pixel i in preprocessed image M_i. After this preprocessing, an unsupervised segmentation network over-segmentations the preprocessed image $\tilde{M}$ into segmentation patches according to the color and spatial information:

$$S = \{S_1, S_2, \ldots, S_K\} = Net_{US}(\tilde{M}), \tag{5.22}$$

where $Net_{US}(\cdot)$ represents the unsupervised segmentation network. The unsupervised segmentation network segments the image into K patches, and the semantic-agnostic patches are denoted as S_k. Since the human body has been segmented from the background, the unsupervised segmentation network can pay more attention to the human body and generate segmentation patches S with clear boundaries. After this, given the segmentation patches $S = \{S_1, S_2, \ldots, S_K\}$ and the weak supervision point-level labels $T = \{T_1, T_2, \ldots, T_N\}$, we define the region-semantic to produce the pixel-level pseudo labels.

Definition of region_semantic. Here, T_n is an annotated semantic point and S_k is a segmentation patch. We define an assignment operator $\Rightarrow$ as the region_semantic strategy. Concretely, when $T_n \subseteq S_k$, and then $S_k = Label(T_n)$. Here, $Label(T_n)$ is the category information of the annotated semantic point T_n. We carry out the region_semantic strategy for all the segmentation patches and then produce the initial pseudo labels $\bar{y}$. However, $K \gg N$, thus there are some segmentation patches that cannot be assigned by the annotated semantic points. In this case, these segmentation patches will be set to 0. Finally, to further improve the accuracy of the initial pseudo labels $\bar{y}$, we optimize it by the box-level labels:

$$y_i = \begin{cases} \bar{y}_i, & i \in B \; and \; \bar{y}_i \neq 0 \\ 0, & i \notin B \\ 255, & i \in B \; and \; \bar{y}_i = 0 \end{cases}, \tag{5.23}$$

where y_i represents the final pseudo labels. Specially, as shown in Eq. (5.14), we set 255 as ignored labels in the final pseudo labels. The ignored labels will not participate in the training process. In this way, region-semantic based unsupervised segmentation strategy can generate high-quality pseudo labels with accurate background labels.

5.4.2 Graph Neural Network Based Weakly Supervised Human Parsing

Weakly supervised human parsing can effectively alleviate the cost of pixel-level annotation, which has attracted the attention of researchers [27–29]. Liu et al. [27] propose to automatically parse the fashion images with weak supervision from the user-generated color-category tags such as "red jeans" and "white T-shirt". They combine the human pose estimation module, the MRF-based color and category inference module and the (super)pixel-level category classifier learning module to generate multiple well-performing category classifiers, which can be directly applied to parse the fashion items in the images. To ease the labeling efforts, Wu et al. [28] propose a novel weakly supervised human parsing method which only requires simple object keypoint annotations for learning. They develop an iterative learning method to generate pseudo part segmentation masks from keypoint labels. With these pseudo masks, they train a FCN network to output pixel-level human parsing predictions. Furthermore, they develop a correlation network to perform joint prediction of part and object segmentation masks and improve the segmentation performance. To overcome the tedious and labor-intensive procedure of annotating pixel-wise masks, Zhao et al. [29] propose a part evolution framework to learn reliable predictions from weak pose annotations, which are much easier to collect. This framework is composed of two essential modules: the first part adaptation module is designed to learn the deep prior knowledge from three related tasks, i.e., pose estimation, part-level and object-level segmentation; the second module is the part evolution module, which refines the part priors from deep predictions with the boundary-aware optimization algorithm. These two modules are conducted iteratively to evolve pose keypoint annotations into reliable part priors. Based on the above analysis, weakly supervised human parsing has achieved some progress in recent years, and graph neural network has been verified effective in processing structured data. Hence, we propose to narrow the gap of pixel-level supervised human parsing and weakly supervised human parsing problem with graph neural network. On the basis of the aforementioned pseudo label generation, we propose an iterative training framework named Dynamic Hierarchical Semantic Graph Neural Network (DHSGNN), which makes the first attempt to parse the human body into fine-grained semantic parts in weakly supervised. In general, the graph data constructed by the human body is inherently flat, and the feature information propagate by the graph neural network. For human body, there are natural hierarchical structure which can be used to distill multiscale features and hierarchical representation by graph pooling operation. In this case, we design a dynamic topology-aware pooling in graph neural network to capture hierarchical structure representation for a better result.

5.4.3 Framework and Procedure

As mentioned above, we propose an iterative training framework named Dynamic Hierarchical Semantic Graph Neural Network (DHSGNN) based on the pixel-level pseudo labels. As shown in Fig. 5.10, this framework is a two-stage learning model which consists of a region-semantic based unsupervised segmentation strategy and a dynamic hierarchical semantic graph neural network(DHSGNN). For DHSGNN, there are three modules: Reliable Pixel Sampling (RPS), Dynamic Topology-aware Pooling (DTP), and Hierarchical Consistency Label Refinement (HCLR). In this chapter, we will detailedly describe the proposed framework DHSGNN. For weakly supervised human parsing, there are three important issues: (1) How to accurately select reliable pixel labels and reduce the negative impact of noisy labels. (2) How to fully exploit semantic structural prior knowledge to capture hierarchical information for weakly supervised network boosting. (3) How to progressively refine the pseudo labels to alleviate the label scarcity problem in weakly supervised human parsing. We design three modules, i.e., RPS, DTP and HCLR, to tackle the three issues mentioned above.

Reliable Pixel Sampling. Since the human parsing framework can achieve a high performance by the pixel-level labels, selecting the reliable pseudo labels is a critical step for the weakly supervised human parsing. There is no way to produce the accurate pseudo labels, and we aim to reduce the impact of noise on model learning as much as possible. To achieve this goal, an intuitive solution is to design an strategy which can sample reliable labels from the pseudo labels. Based on this, we propose a reliable pixel sampling (RPS) strategy to alleviate the negative impact of noisy labels. Concretely, RPS defines the following confident constraint. We hypothesize that the reliable pixels depend on the consistency of the high-confidence prediction with the

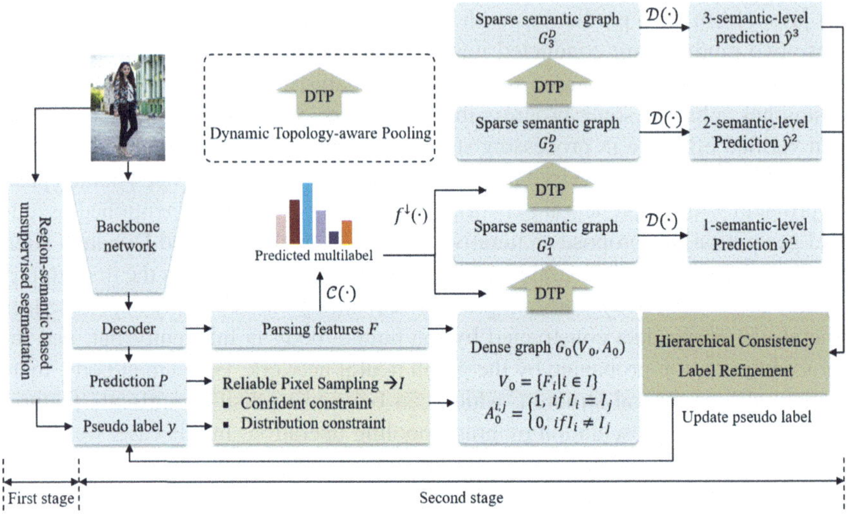

Fig. 5.10 Diagram of the two-stage framework to predict reliable fine-grained human parsing results from weakly supervised box-level and point-level annotations

pseudo labels, and we name it a confident constraint. In detail, we judge one pixel is a reliable pixel when the annotation of the pseudo label is the same as the initial predicted result, as follows:

$$\begin{aligned} \check{y} &= Net_{backbone}(M, y), \\ I &= \left\{ i | index(\max(\check{y}_i)) = y_i \right\}, \end{aligned} \tag{5.24}$$

where $\check{y}$ is the initial predicted results produced by the backbone network $Net_{backbone}(\cdot)$ supervised by the initial pseudo labels y. The $index(\cdot)$ represents the category of the maximum value in the predicted vector $\check{y}_i$, and I is the set of sampled reliable pixels. In practice, we find that this strict confident constraint can sample the reliable pixels and efficiently remove the noisy labels. However, this strict confident constraint also improperly deletes some correct labels in pseudo labels. To improve the accuracy of the proposed confident constraint, we relax the confident constraint to encourage retaining the pixels of pseudo labels as much as possible. Then, the Eq. (5.15) is optimized into the following:

$$\begin{aligned} \check{y}^{top_3} &= index(rank^{top_3}(\check{y})), \\ I &= \left\{ i | y_i \in \check{y}_i^{top_3} \right\}. \end{aligned} \tag{5.25}$$

As we can see, instead of only selecting the max value of the initial predicted result $\check{y}$, we rank the $\check{y}$ and select the top_3 with the higher confidence as candidates. The $\check{y}^{top_3}$ are the candidates, and the pixel i will be judged as a reliable pixel if the corresponding pseudo label y_i is included in $\check{y}_i^{top_3}$. We attempt to improve the robustness of the model by relaxing the search space, and the top_3 confidences are verified to be the most effective parameter in practice. For example, a pixel i of the right arm, although the annotation in the pseudo label is accurate, it is hard for the backbone network to precisely predict this pixel, i.e., the pixel i of the right arm could be predicted as the left arm for similar features. However, when the search space is relaxed to top_3, both the right arm and left arm will be included in the relaxed search space $\check{y}_i^{top_3}$, and then this indeterminate pixel could be selected out. By relaxing the confidence constraint, we can select as many reliable pixels as possible and improve the robustness of the model.

Construct Reliable Graph for Human Parsing To improve the accuracy of constructed graph while ensuring that the graph neural network can extract accurate global structural features, we only build the graph based on reliable pixels obtained through RPS strategy. Specifically, we treat the sampled reliable pixels as graph nodes and initialize the feature vectors of these nodes using the corresponding features obtained from the backbone network:

$$V_0 = \{F_i | i \in I\}, \tag{5.26}$$

where I is the set of reliable pixels, F represents the extracted high-dimensional feature of the backbone, and F_i stands for the feature vector of a reliable pixel. All of the reliable pixels will form the dense graph nodes $V_0 \in \mathbb{R}^{d \times C}$ based on the high-dimensional feature F. After this, the dense adjacency matrix A_0 is computed by the pseudo labels obtained through RPS strategy:

$$A_0^{i,j} = \begin{cases} 1, & if\ I_i = I_j \\ 0, & if\ I_i \neq I_j, \end{cases} \tag{5.27}$$

where I_i and I_j are the pseudo labels of reliable pixels i and j respectively. The generated V_0 and A_0 instantiate the pixel-level dense graph $G_0(V_0, A_0)$. Since we only apply the reliable pixels to construct the dense graph, it can guarantee the precision of G_0 and suppress the negative impact of noisy labels. Furthermore, to obtain global structural features of the human body, we apply graph pooling to coarsen the constructed graph G_0, resulting in the previously defined semantic graph focused on the human body. Graph pooling is a technique used to reduce the graph's size while preserving essential structural information. By applying graph pooling, we aim to capture high-level patterns and features that represent the overall structure of the human body. This semantic graph can then be used for various tasks, such as semantic segmentation or pose estimation, to better understand and analyze human body structures and relationships:

$$G_s = Graphpooling(G_0) \in \mathbb{R}^{N \times d}. \tag{5.28}$$

In order to obtain highly precise and detailed boundaries of human body, we continue to pool the semantic graph into a coarser grained graph, which constructed only by foreground and background:

$$G_{f\&g} = Graphpooling(G_s) \in \mathbb{R}^{2\times d}. \tag{5.29}$$

G_s and $G_{f\&g}$ form a multi-granularity hierarchical semantic graph focused on the human body. This multi-granularity hierarchical semantic graph allows us to analyze the human body at different levels of granularity, from coarse to fine. It incorporates both the global structural features obtained from the semantic graph G_s and the precise edge features derived from the $G_{f\&g}$. As a result, the multi-granularity hierarchical semantic graph provides a comprehensive and detailed representation of the human body, making it producing a high performance model. Finally, we apply a decoder model to convert the graph features into human parsing predicted results:

$$\begin{aligned} F_s &= \Phi(V_s, W_n), \\ \hat{y}_{fine} &= \mathcal{D}(conv(cat(F, F_s))), \end{aligned} \tag{5.30}$$

and coarse human parsing results:

$$\begin{aligned} F_{f\&g} &= \Phi(V_{f\&g}, W_n), \\ \hat{y}_{coarse} &= \mathcal{D}(conv(cat(F, F_{f\&g}))). \end{aligned} \tag{5.31}$$

Hierarchical Consistency Label Refinement As described in Sect. 3, our aim is to generate precise pseudo labels by marking many challenging pixels as ignored labels (255). Although these pixels are not included in the training process, their impact should not be disregarded, and an effective relabeling method for this data can yield significant benefits. Consequently, leveraging the outputs of DHSGNN, we introduce a hierarchical consistency label refinement (HCLR) strategy to reassign labels to these ignored pixels. Specifically, we begin by mapping the predictions at the G_s to the $G_{f\&g}$ based on the inherent structure of the human body:

$$\hat{y}_{fine,coarse} = f^{\downarrow}(\hat{y}_{fine}), \tag{5.32}$$

where $f^{\downarrow}(\cdot)$ is the projection function, i.e., $\{face, hat, hair, sunglasses, etc.\} \overset{f^{\downarrow}(\cdot)}{\rightarrow} Upper\text{-}body$, $\{socks, pants, skirt, etc.\} \overset{f^{\downarrow}(\cdot)}{\rightarrow} Lower\text{-}body$. Then, the consistency between $\hat{y}_{fine,coarse}$ and $\hat{y}_{coarse}$ is defined as:

$$\hat{y}_{coarse,i} = eq(\hat{y}_{fine,coarse,i}, \hat{y}_{coarse,i}) = \begin{cases} \hat{y}_{coarse,i}, & \hat{y}_{fine,coarse,i} = \hat{y}_{coarse,i} \\ 255, & \hat{y}_{fine,coarse,i} \neq \hat{y}_{coarse,i}, \end{cases} \tag{5.33}$$

where i is the pixel index, and $\hat{y}_{coarse,i} \in \{0, 1, 2, 255\}$. With the hierarchical consistent representation $\hat{y}_{coarse,i}$, we update the $\hat{y}_{fine}$ as follows:

$$\hat{y}_{fine} = \langle \hat{y}_{coarse,i}, \hat{y}_{fine,i} \rangle = \begin{cases} \hat{y}_{fine,i}, & \hat{y}_{coarse,i} = 0, 1 \\ 255, & \hat{y}_{coarse,i} = 255. \end{cases} \tag{5.34}$$

Finally, the pseudo label is refined by $\hat{y}_{fine}$:

$$y_i^{up} = \begin{cases} y_i, & y_i \neq 255 \\ \hat{y}_{fine}, & y_i = 255. \end{cases} \tag{5.35}$$

This strategy effectively addresses the issue of limited labels in weakly supervised human parsing by leveraging the online updated pseudo labels. By incorporating these updated pseudo labels, we are able to overcome the scarcity of labeled data, leading to improved performance in the task of human parsing.

5.4.4 Experimental Results and Discussions

The Effect of Pseudo Label. In this chapter, we propose a novel approach for generating pixel-level pseudo labels using both point-level labels and box-level labels, instead of relying on the commonly used Class Activation Maps (CAM) method. Our method involves a region-semantic based unsupervised segmentation strategy, which segments the input image into patches. Each patch is then extended to cover the corresponding semantic annotated point. We refer to this strategy as a region-semantic based unsupervised segmentation. To evaluate the effectiveness of our pseudo label generation method, we compare it with the popular super-pixel method called SLIC. The qualitative results of this comparison can be observed in Fig. 5.11. Our designed strategy demonstrates the ability to generate relatively accurate pseudo labels, even for complex postures, as evident from the first row of Fig. 5.11. We argue that these high-quality pseudo labels can further enhance the performance of our parsing results. However, when we employ the SLIC method for image segmentation, none of the three segmentation schemes (SLIC/50, SLIC/100, SLIC/150) produce satisfactory results. Additionally, we investigate the impact of adding extra annotated semantic points on model performance. Table 5.7 presents the results for each semantic category when adding 1–5 annotated semantic points. Notably, using only one semantic point still achieves favorable performance with 92.75% pixel accuracy and 57.05% mIoU on the ATR dataset. This result is likely due to the effectiveness of our designed region-semantic based unsupervised segmentation strategy. To verify our argument, we visualize the pseudo labels generated using different numbers of semantic points in Fig. 5.12. Point = 1 indicates that each semantic part is labeled by only one point, the quality of the pseudo label is comparable to point = 2 and point = 4. Moreover, even when point=1, the small parts (such as sunglasses and scarf) can still be accurately labeled, as indicated by the white dashed rectangle in Fig. 5.12. Based on the quantitative and qualitative analysis of our experimental results, we set point = 1 to strike a balance between labeling costs and model performance in this chapter.

Fig. 5.11 Qualitative results of region-semantic based unsupervised segmentation strategy and SLIC method on ATR test [30]. **a** The pixel-level pseudo labels of region-semantic based unsupervised segmentation strategy. **b** The pixel-level pseudo labels of SLIC method with segmenting the input images into 50 patches (denoted as SLIC/50). **c** The pixel-level pseudo labels of SLIC method with segmenting the input images into 100 patches (denoted as SLIC/100). **d** The pixel-level pseudo labels of SLIC method with segmenting the input images into 150 patches (denoted as SLIC/150)

Table 5.7 Comparison of pseudo labels produced by various number points on ATR test [30]

Number points/category	Pixel accuracy	mIoU
1	92.75	57.05
2	92.44	56.14
3	92.56	56.47
4	92.71	**57.50**
5	**92.80**	56.73

Comparisons with the state-of-the-art methods on Single-person part parsing dataset. Due to the novelty of this work which attempts to segment the human body into fine-grained parts under weak supervision, we compared our proposed algorithm with a fully supervised method trained using pixel-level ground truth, as shown in Table 5.8. Our proposed algorithm only utilizes weakly supervised information during the training phase and does not require any additional supervision during the testing phase. LIP (Look Into Person) is currently the largest single-human parsing dataset and is commonly used as a standard benchmark for evaluating algorithm performance. It is worth noting that with only about 18 points and a bounding box as supervision, the proposed method achieves an average accuracy of 57.16%, outperforming all reported fully supervised methods. Additionally, the proposed method does not separate dress, scarf, and skirt. Dress and scarf are small objects in the LIP dataset, which are also challenging to segment in other fully supervised meth-

Fig. 5.12 Visualization of pseudo labels produced by various number of points

ods. Therefore, under weak supervision, the proposed method faces difficulties in recognizing and segmenting them accurately. Furthermore, we visualize some representative samples in Fig. 5.13 to explore the reasons for the segmentation errors in skirts. As observed within the white dashed box, the skirt is misinterpreted as pants. In the LIP dataset, the annotations for skirts are often smaller compared to those for pants. Therefore, using weak supervision alone, the framework may struggle to capture feasible and effective features to distinguish skirt regions accurately. Especially when the semantic structure of skirts and pants is located in the same position on the human body, the ambiguity arising from their overlapping semantic structure makes it challenging to achieve accurate segmentation results solely under weak supervision.

Comparisons with the state-of-the-art methods on Multi-person part parsing dataset. Furthermore, we evaluated the performance of the proposed algorithm on the CIHP dataset. In CIHP, each image contains approximately three individuals, and all individuals are annotated with 19 semantic labels. Additionally, the dataset includes challenging poses and mutual occlusions, resulting in limited visibility of the whole body. Under such circumstances, accurately parsing the CIHP dataset under weak supervision alone is challenging, and there has been no specific weakly supervised research targeting the CIHP dataset. In this context, we compared the fully supervised method with the proposed weakly supervised method. We employed ResNet-101 as the backbone and used both bounding box-level labels and point-level labels as supervised information. As shown in Table 5.9, the baseline model achieves an mIoU of 27.55% under weak supervision and 40.95% under strong

Table 5.8 Comparison of mIoU and mean accuracy on LIP validation set [5]

Methods	bkg	Hat	Hair	Glove	s-glass	u-clot	Dress	Coat	Socks	Pants	j-suit
SegNet [31]	70.62	26.60	44.01	0.01	0.00	34.46	0.00	15.97	3.59	33.56	0.01
FCN-8 s [32]	78.02	39.79	58.96	5.32	3.08	49.08	12.36	26.82	15.66	49.41	6.48
DeepLab (VGG-16) [13]	83.25	57.94	66.11	23.32	18.40	60.94	23.17	47.03	34.51	64.00	22.38
Attention [16]	84.00	58.87	**66.78**	23.32	19.48	63.20	29.63	49.70	35.23	66.04	24.73
DeepLab (ResNet-101) [13]	**84.09**	**59.76**	66.22	**28.76**	23.91	**64.95**	**33.68**	**52.86**	**37.67**	**68.05**	**26.15**
DHSGNN	77.33	47.43	59.49	20.67	**24.46**	54.56	10.10	41.57	34.05	57.00	0.00
Methods	Scarf	Skirt	Face	l-arm	r-arm	l-leg	r-leg	l-shoe	r-shoe	mIoU	Mean accuracy
SegNet [31]	0.00	0.00	52.38	15.30	24.23	13.82	13.17	9.26	6.47	18.17	24.00
FCN-8 s [32]	0.00	2.16	62.65	29.78	36.63	28.12	26.05	17.76	17.70	28.29	36.75
DeepLab (VGG-16) [13]	14.29	18.74	69.70	49.44	51.66	37.49	34.60	28.22	22.41	41.64	51.64
Attention [16]	12.84	20.41	**70.58**	50.17	**54.03**	38.35	37.70	26.20	27.09	42.92	54.39
DeepLab (ResNet-101) [13]	**17.44**	**25.23**	70.00	**50.42**	53.89	39.36	38.27	26.95	28.36	**44.80**	55.62
DHSGNN	0.00	0.00	66.98	42.30	42.48	**42.22**	**43.64**	**29.43**	**28.81**	36.12	**57.16**

Fig. 5.13 Visualization of some failure cases for skirt on the LIP validation set

supervision. The substantial gap between weak supervision and strong supervision confirms that achieving high performance in fine-grained multi-person part analysis heavily relies on pixel-level ground truth. However, when using the proposed model for multi-person part parsing, we observe significant improvement with an mIoU of 33.75%. This improvement can be attributed to the proposed algorithm, which can mitigate the negative impact of noisy labels and iteratively improve the pseudo-labels. To further validate the effectiveness of the proposed algorithm, we added 10,000 pixel-level ground truth samples from the CIHP training set and denoted it as DHSGNN+. We can observe that DHSGNN+ produces results comparable to the fully supervised baseline model. Due to the considerable challenges in the CIHP dataset, there still exists a significant gap between the results of our proposed weakly supervised algorithm and existing fully supervised algorithms. Nevertheless, the reported quantitative results validate the feasibility and effectiveness of the proposed algorithm for parsing multi-person under weak supervision.

Table 5.9 Comparison of mIoU on CIHP validation set [20]. The symbol † marks that the experimental results are reproduced by ourselves

Methods	Anno.	mIoU
Deeplab (ResNet-101) [13]†	Fully-supervised	40.95
PGN [20]	Fully-supervised	55.80
Graphonomy [8]	Fully-supervised	57.78
Deeplab (ResNet-101) [13]†	Box&point	27.55
DHSGNN	Box&point	33.75
DHSGNN+	Box&point&10000fully-supervised	40.96

In this chapter, we introduce two different ways of constructing graphs for human parsing: the spatial graph and the semantic graph. Based on these, we propose the Context Prior-based Semantic-Spatial Graph Network (CP-SSGNet), which leverages the multi-level structural context prior information of the human body to design optimal graph modeling for human parsing. In this model, we propose three modules, namely SCM, SPM, and IAM. Experiments demonstrate that CP-SSGNet can effectively address hard samples and reduce local consistency errors and semantic structure errors. Further, we try to explore the possibility of graph neural network model in weakly supervised human parsing. With box-level labels and point-level labels, we propose a two-stage learning framework that combines pseudo label generation with a graph neural network-based weakly supervised human parsing. This framework includes a region-semantic-based unsupervised segmentation strategy and a Dynamic Hierarchical Semantic Graph Neural Network (DHSGNN). DHSGNN comprises three modules: RPS, DTP, and HCLR. Experiments under various settings demonstrate that the proposed DHSGNN achieves results comparable to some strongly supervised methods. However, considering that the two-stage framework cannot be integrated into a single model, we aim to simplify the model using weak pose annotations to further improve the performance of weakly supervised human parsing.

References

1. Li, R., Wang, S., Zhu, F., Huang, J.: Adaptive graph convolutional neural networks. In: Proceedings of the AAAI Conference on Artificial Intelligence, pp. 3546–3553 (2018)
2. Dai, H., Kozareva, Z., Dai, B., Smola, A.J., Song, L.: Learning steady-states of iterative algorithms over graphs. In: International Conference on Machine Learning, vol. 80, pp. 1114–1122 (2018)
3. Wu, Y., Bourahla, O.E.F., Li, X., Wu, F., Tian, Q., Zhou, X.: Adaptive graph representation learning for video person re-identification. IEEE Transactions on Image Processing 29, 8821–8830 (2020)
4. Shi, L., Zhang, Y., Cheng, J., Lu, H.: Skeleton-based action recognition with multi-stream adaptive graph convolutional networks. IEEE Transactions on Image Processing 29, 9532–9545 (2020)

5. Liang, X., Gong, K., Shen, X., Lin, L.: Look into person: joint body parsing & pose estimation network and a new benchmark. IEEE Transactions on pattern analysis and machine intelligence 41(4), 871–885 (2019)
6. Zhang, Z., Su, C., Zheng, L., Xie, X.: Correlating edge, pose with parsing. In: Proceedings of the IEEE Conference on Computer Vision and Pattern Recognition, pp. 8897–8906 (2020)
7. He, H., Zhang, J., Zhang, Q., Tao, D.: Grapy-ml: graph pyramid mutual learning for cross-dataset human parsing. In: Proceedings of the AAAI Conference on Artificial Intelligence, pp. 10949–10956 (2020)
8. Gong, K., Gao, Y., Liang, X., Shen, X., Wang, M., Lin, L.: Graphonomy: universal human parsing via graph transfer learning. In: Proceedings of the IEEE Conference on Computer Vision and Pattern Recognition, pp. 7450–7459 (2019)
9. Wang, W., Zhou, T., Qi, S., Shen, J., Zhu, S.: Hierarchical human semantic parsing with comprehensive part-relation modeling. IEEE Transactions on pattern analysis and machine intelligence 44(7), 3508–3522 (2022)
10. Li, X., Wang, W., Hu, X., Yang, J.: Selective kernel networks. In: Proceedings of the IEEE Conference on Computer Vision and Pattern Recognition, pp. 510–519 (2019)
11. Xie, S., Girshick, R.B., Dollár, P., Tu, Z., He, K.: Aggregated residual transformations for deep neural networks. In: Proceedings of the IEEE Conference on Computer Vision and Pattern Recognition, pp. 5987–5995 (2017)
12. Krizhevsky, A., Sutskever, I., Hinton, G.E.: Imagenet classification with deep convolutional neural networks. In: Neural Information Processing Systems, pp. 1106–1114 (2012)
13. Chen, L., Papandreou, G., Kokkinos, I., Murphy, K., Yuille, A.L.: Deeplab: semantic image segmentation with deep convolutional nets, atrous convolution, and fully connected crfs. IEEE Transactions on pattern analysis and machine intelligence 40(4), 834–848 (2018)
14. Szegedy, C., Vanhoucke, V., Ioffe, S., Shlens, J., Wojna, Z.: Rethinking the inception architecture for computer vision. In: Proceedings of the IEEE Conference on Computer Vision and Pattern Recognition, pp. 2818–2826 (2016)
15. Szegedy, C., Ioffe, S., Vanhoucke, V., Alemi, A.A.: Inception-v4, inception-resnet and the impact of residual connections on learning. In: Proceedings of the AAAI Conference on Artificial Intelligence, pp. 4278–4284 (2017)
16. Chen, L., Yang, Y., Wang, J., Xu, W., Yuille, A.L.: Attention to scale: scale-aware semantic image segmentation. In: Proceedings of the IEEE Conference on Computer Vision and Pattern Recognition, pp. 3640–3649 (2016)
17. Xia, F., Wang, P., Chen, X., Yuille, A.L.: Joint multi-person pose estimation and semantic part segmentation. In: Proceedings of the IEEE Conference on Computer Vision and Pattern Recognition, pp. 6080–6089 (2017)
18. Luo, Y., Zheng, Z., Zheng, L., Guan, T., Yu, J., Yang, Y.: Macro-micro adversarial network for human parsing. In: Proceedings of the European Conference on Computer Vision, vol. 11213, pp. 424–440
19. Fang, H., Lu, G., Fang, X., Xie, J., Tai, Y., Lu, C.: Weakly and semi supervised human body part parsing via pose-guided knowledge transfer. In: Proceedings of the IEEE Conference on Computer Vision and Pattern Recognition, pp. 70–78 (2018)
20. Gong, K., Liang, X., Li, Y., Chen, Y., Yang, M., Lin, L.: Instance-level human parsing via part grouping network. In: Proceedings of the European Conference on Computer Vision, vol. 11208, pp. 805–822 (2018)
21. Chen, L., Zhu, Y., Papandreou, G., Schroff, F., Adam, H.: Encoder-decoder with atrous separable convolution for semantic image segmentation. In: Proceedings of the European Conference on Computer Vision, vol. 11211, pp. 833–851 (2018)
22. Wang, W., Zhang, Z., Qi, S., Shen, J., Pang, Y., Shao, L.: Learning compositional neural information fusion for human parsing. In: IEEE International Conference on Computer Vision, pp. 5702–5712 (2019)
23. Li, P., Xu, Y., Wei, Y., Yang, Y.: Self-correction for human parsing. IEEE Transactions on pattern analysis and machine intelligence 44(6), 3260–3271 (2022)

24. Nie, X., Feng, J., Yan, S.: Mutual learning to adapt for joint human parsing and pose estimation. In: Proceedings of the European Conference on Computer Vision, vol. 11209, pp. 519–534 (2018)
25. Ruan, T., Liu, T., Huang, Z., Wei, Y., Wei, S., Zhao, Y.: Devil in the details: towards accurate single and multiple human parsing. In: Proceedings of the AAAI Conference on Artificial Intelligence, pp. 4814–4821 (2019)
26. Zhang, Z., Su, C., Zheng, L., Xie, X.: Correlating edge, pose with parsing. In: Proceedings of the IEEE Conference on Computer Vision and Pattern Recognition, pp. 8897–8906 (2020)
27. Liu, S., Feng, J., Domokos, C., Xu, H., Huang, J., Hu, Z., Yan, S.: Fashion parsing with weak color-category labels. IEEE Transactions on Multimedia 16(1), 253–265 (2014)
28. Wu, Z., Lin, G., Cai, J.: Keypoint based weakly supervised human parsing. Image and Vision Computing 91 (2019)
29. Zhao, Y., Li, J., Zhang, Y., Tian, Y.: From pose to part: weakly-supervised pose evolution for human part segmentation. IEEE Transactions on pattern analysis and machine intelligence 45(3), 3107–3120 (2023)
30. Liang, X., Xu, C., Shen, X., Yang, J., Liu, S., Tang, J., Lin, L., Yan, S.: Human parsing with contextualized convolutional neural network. In: IEEE International Conference on Computer Vision, pp. 1386–1394 (2015)
31. Badrinarayanan, V., Kendall, A., Cipolla, R.: Segnet: a deep convolutional encoder-decoder architecture for image segmentation. IEEE Transactions on pattern analysis and machine intelligence 39(12), 2481–2495 (2017)
32. Long, J., Shelhamer, E., Darrell, T.: Fully convolutional networks for semantic segmentation. In: Proceedings of the IEEE Conference on Computer Vision and Pattern Recognition, pp. 3431–3440 (2015)

Chapter 6
Dynamic Graph Neural Networks for Human Parsing

6.1 Introduction

Graphs are prevalent in various real-world applications, and Graph Convolutional Networks (GCNs) have gained great popularity in tackling various analytics tasks on graph and network data. However, some recent studies raise concerns about whether GCNs can optimally integrate node features and topological structures in a complex graph with rich information. The conventional graph remains static, meaning that both the graph structure and the node features remain fixed over time. The adaptive graph structure allows the framework to effectively handle the evolving entities and relations, making it suitable for scenarios where the underlying data is subject to changes and updates over time. By continuously updating the graph based on the evolving information, the framework remains adaptable and capable of capturing the most relevant interactions and relationships among entities. As a result, it can maintain accurate and up-to-date representations of the underlying data, enabling it to perform effectively in dynamic environments. Over the years, various machine learning models have been developed to tackle the challenges of dynamic graphs, capturing their evolving structure and changes over time. Li et al. [1] proposed a generalized and flexible graph CNN which is designed to handle data with arbitrary graph structures. The key feature of this approach is the ability to learn a task-driven adaptive graph for each graph data instance during the training process. And to achieve efficient graph learning, the approach introduces a new distance metric learning to obtain the optimal distance metric parameters shared among the data. Wang et al. [2] conduct large experiments and clearly show that the capability of the state-of-the-art GCNs in fusing node features and topological structures is distant from optimal or even satisfactory. In order to remedy the weakness and design a new type of GCNs, they propose an Adaptive Multi-channel Graph Convolutional Networks for semi-supervised classification (AM-GCN). The central idea is that they extract the specific and common embeddings from node features, topological structures, and their combinations simultaneously, and use the attention mechanism

W. Liu et al., *Graph Neural Network Methods and Applications in Scene Understanding*,
https://doi.org/10.1007/978-981-97-9933-6_6

to learn adaptive importance weights of the embeddings. In the skeleton-based action recognition, GCNs have demonstrated remarkable performance. These GCNs models human body skeletons as spatiotemporal graphs, where each joint of the skeleton represents a node in the graph, and the edges between the nodes capture the spatial and temporal relationships between the joints. By leveraging the graph structure of the skeleton data, GCNs can effectively capture the dependencies and interactions between different body joints, which are crucial for understanding human actions. In existing GCN-based methods, the graph topology is typically set manually and remains fixed for all layers and input samples. This fixed graph structure may not be optimal for tasks that involve hierarchical GCNs and diverse samples, such as action recognition. In this case, Shi et al. [3] propose a novel two-stream Adaptive Graph Convolutional Network (2s-AGCN) for skeletonbased action recognition. The topology of the graph in this model can be either uniformly or individually learned in an end-to-end manner. This data-driven method increases the flexibility of the model for graph construction and brings more generality to adapt to various data samples. Moreover, a two-stream framework is proposed to model both the first-order and the second-order information simultaneously, which shows notable improvement for the recognition accuracy. In this chapter, we focus on modelling an adaptive graph for human parsing task and boost the performance by the dynamic graph neural networks.

6.2 Dynamic Graph Construction for Human Parsing

The rise of GCNs has obtained significant attention for graph-based methods, as they offer optimal graph reasoning and enable the incorporation of long-range context information. In human parsing, graph-based methods [4, 5] generally directly encode the topological structure into a fixed graph representation and subsequently leverage GCNs to facilitate the propagation of information throughout the fixed graph. However, it is common that a training image sample contains multiple classes, but not all of the labeled classes in the dataset are present in the image. For instance, the legs in an image might disappear due to occlusion, or certain classes like sunglasses or hats may not be present in the image, as depicted in Fig. 6.1. Figure 6.1a provides a schematic representation of a fixed graph commonly employed in existing methods. In contrast, Fig. 6.1b depicts the adaptive graph, which has been demonstrated to be more accurate and optimal for this specific training sample. It becomes evident that the fixed graph is no longer an ideal choice for accommodating the diversity of data in human parsing tasks. As a result, to overcome the limitations of the fixed graph, we are making efforts to design an adaptive graph. However, the generation of an adaptive graph for each training sample remains a challenge for current methods. In light of this, this chapter aims to propose a novel framework that can effectively generate an adaptive graph for human parsing while preserving the advantages of the

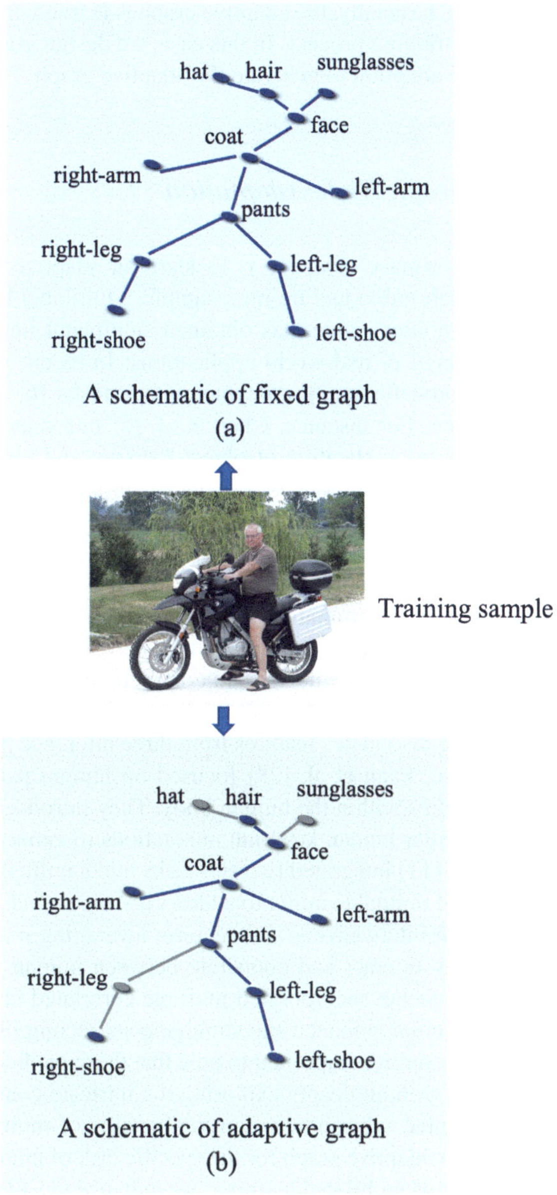

Fig. 6.1 **a** A schematic of fixed graph. **b** A schematic of adaptive graph. Gray nodes denote the labels that do not appear in the training sample

fixed graph. Crucially, the adaptive graph is learned uniquely for each image sample during the training process. In this case, we design graph node adaptation and graph connection adaption to generate the adaptive graph.

6.2.1 Graph Node Adaptation

Here, our primary objective is to learn an adaptive graph structure uniquely tailored to each individual training sample. Multilabel learning, being a fundamental task in computer vision, has obtained significant interest from researchers due to its wide range of real-world applications. In recent years, multi-task learning has been explored for many computer vision tasks [6, 7] to improve generalization performance. For instance, Chen et al. [6] introduced an innovative pipeline that leverages joint application of object tracking and video object segmentation tasks to amplify segmentation performance. Similarly, Cao et al. [7] pursued the joint optimization of object detection and video instance segmentation, employing a mask alignment weighting loss to promote mutual benefit between these interrelated tasks. Furthermore, to capture the rich structure features, some previous works [8, 9] utilized human pose estimation to accomplish the human parsing. For instance, Wang et al. [10] proposed to combine neural networks with the compositional hierarchy of human bodies for efficient and complete human parsing. This method explicitly models the compositional and decompositional relations in human bodies, respectively, and consolidates features from three inference processes across the hierarchy. Furthermore, Fang et al. [28] focused on human parsing by leveraging anatomical similarities within the human body. They introduced an innovative strategy that utilized similar human keypoint annotations to generate parsing results. Moreover, Gong et al. [11] integrated two sub-tasks into a unified network, allowing for shared features and mutual learning to address instance-level human parsing. Nie et al. [12] designed an interweaving architecture, leveraging mutual guidance information to dynamically interact and cooperate between human parsing and pose estimation. These approaches incorporated multiple correlated tasks into a unified framework, enabling mutual guidance and simultaneous accomplishment of multiple prediction tasks. However, it's important to note that these methods primarily fused knowledge expression without deeply exploring the intricate connections between these tasks. In this chapter, we capitalize on the benefits of multilabel learning to aid in constructing an adaptive graph for the specific task of human parsing. By leveraging the capabilities of multilabel learning, we enhance the adaptability and effectiveness of the graph structure, leading to improved performance and representation learning in human parsing tasks. In our approach, we utilize the ResNet-101 [13] as the backbone network, along with the context encoding module (PSP) [14]. The output features X from this backbone are subsequently used to predict the multilabel. Inspired by previous work, we formally define the multilabel learning task as a mapping function with two-layer MLP and a nonlinear activation function:

$$M = sigmoid(\mathcal{F}(GMP(X))), \tag{6.1}$$

where $X \in \mathbb{R}^{C \times H \times W}$ is the output features of the backbone, $M \in \mathbb{R}^{1 \times N}$ is the estimating multilabel classification scores for the input images, and N is the number of the labels. $GMP(\cdot)$ is the global max pooling, and $\mathcal{F}$ is the two-layer MLP. Then, we compute the cross-correlation matrix based on probability score:

$$\mathbb{Q} = M^T M, \tag{6.2}$$

here, we define $\mathbb{Q} \in \mathbb{R}^{N \times N}$ as the adaptive matrix. After applying the *sigmoid* function, the values of M and $\mathbb{Q}$ are transformed to the range of 0 to 1.

In our method, we leverage $\mathbb{Q}$ for both graph node adaptation and graph connection adaptation, allowing the model to dynamically delete nodes and edges. To achieve this adaptability effectively, the values of $\mathbb{Q}$ require further optimization. Ideally, smaller values of $\mathbb{Q}$ are close to 0, indicating nodes and edges that need to be discarded, while larger values are close to 1, suggesting nodes and edges that should be retained in the adaptive graph. By appropriately optimizing the values of $\mathbb{Q}$ within this range, the proposed method can accurately and adaptively adjust the graph structure, resulting in improved representation learning and enhanced performance in human parsing. In this case, we define a 0–1 polarization function to optimize the adaptive matrix $\mathbb{Q}$ as follows:

$$Q = \mathbb{E}(\mathbb{Q}) = \begin{cases} v^2, & 0 < v \leqslant \beta \\ e^{-}(v-1)^2, & \beta < v < 1, \end{cases} \tag{6.3}$$

where $\mathbb{E}(\cdot)$ is the 0–1 polarization function, v is the value in adaptive matrix $\mathbb{Q}$, β is a threshold parameter. With this optimization function, the values in the adaptive matrix Q are distributed around 0 and 1.

As mentioned in Sect. 5.2.2, we define the fixed undirected semantic graph based on the human body knowledge prior, denoted as $G = (V, E, A)$, where V, E, A are the set of nodes, set of edges and adjacency matrix. $|V| = N$ is the number of the labeled semantic parts in dataset. E denotes the fixed connected relationship for all N nodes. $A \in \mathbb{R}^{N \times N}$ is the adjacency matrix with N nodes. Specifically, $A_{i,j} = 1$ denotes there is an edge between nodes i and j, otherwise, $A_{i,j} = 0$. It is noteworthy that the construction of the fixed graph relies on the inherent topological physical structure of the human body, and all the connection relationships are represented in the adjacency matrix A. The visualized schematic of fixed graph is illustrated in Fig. 6.1a. As shown in Fig. 6.1b, it is a common occurrence that a training image sample contains multiple classes, but not all of the labeled classes. In order to improve representation learning, we aim to explore a solution, named Graph Node Adaptation (GNA), which can adaptively activate the graph nodes for human parsing. Indeed, a crucial aspect of GNA lies in its capability to learn an exclusive graph for each image sample during the training process. This adaptability ensures that the graph structure is tailored to the specific characteristics and context of each image, contributing to more effective and accurate representation learning in the task of human parsing. Indeed, the primary goal of GNA is to adaptively remove nodes that may correspond

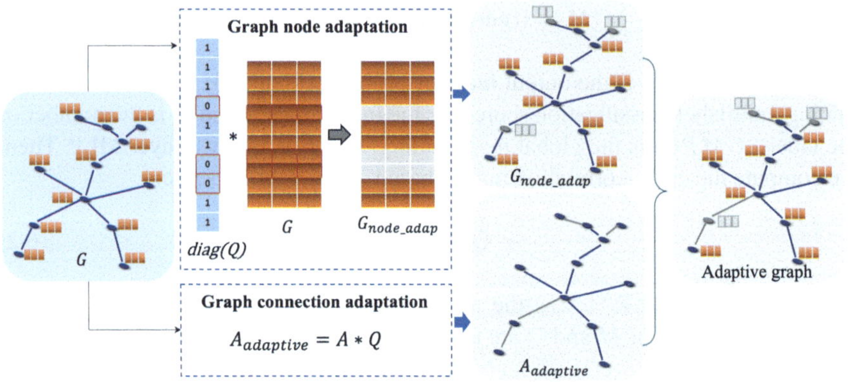

Fig. 6.2 Illustration of graph node adaption and graph connection adaption. The gray nodes denote the nodes which are adaptively removed entire features of nodes that may include noise. The gray lines denote the removed connections

to noise nodes, as illustrated in Fig. 6.2. Exactly, the GNA approach facilitates each node to dynamically propagate with varying subsets of neighbors across different training samples. This adaptive propagation mechanism allows nodes to selectively interact with relevant neighbors, tailored to the specific characteristics and context of each training sample. Consequently, GNA enhances the ability of the model to capture meaningful and discriminative information from the data, ultimately leading to improved performance and generalization. Concretely, for the fixed graph G, the entire feature vectors of some nodes are adaptively removed with the predicted probability of multilabel learning, i.e., some rows of G are set to $\mathbf{0}$, as depicted by gray in Fig. 6.2. To achieve this, we utilize the learned adaptive matrix Q and calculate the graph node features as follows:

$$G_{node\_adap} = diag(Q) * G, \tag{6.4}$$

where $diag(Q)$ represents the diagonal values of the adaptive matrix Q, the symbol $*$ denotes dot products, and G_{node_adap} signifies the graph generated through adaptive optimization. Using this function, the feature of node i is updated by its feature vector with its corresponding adaptive matrix value, denoted as $G^i_{node\_adap} = Q_{i,i} * G^i$. Notably, when the value of $Q_{i,i}$ is close to 0, the entire feature vector of node i is effectively forced to be set to $\mathbf{0}$, as visually represented by the color gray in Fig. 6.2.

Based on the aforementioned analysis, GNA incorporates an adaptive pooling strategy, which allows for the selective removal of certain graph nodes from the fixed graph based on the content of the image. The GNA enables the model to facilitate each node's adaptive propagation with different subsets of neighbors across various training samples, leading to improved graph reasoning and the extraction of high-level features. Moreover, GNA effectively employs this pooling strategy to enhance model robustness and mitigate potential issues like over-smoothing and overfitting, which can arise during the training process. By dynamically adjusting the graph structure and feature propagation, GNA ensures that the model efficiently captures relevant information while mitigating the risks associated with excessively smoothing the representations. As a result, model equipped with GNA achieves improved performance and generalization capabilities.

6.2.2 Graph Connection Adaption

After the GNA, the graph G_{node_adap} is fed into graph convolutional layer for high-level feature extraction:

$$H_{l+1} = \sigma(\widetilde{D}^{-\frac{1}{2}}\widetilde{A}\widetilde{D}^{-\frac{1}{2}}H_l W_l), \tag{6.5}$$

where the adjacency matrix $\widetilde{A} = A + I$ is obtained by the original adjacency matrix A and the identity matrix I. $\widetilde{D}_{i,i} = \sum_j(\widetilde{A}_{i,j})$ is a diagonal matrix of node degrees, W_l is a learnable parameter matrix, $\sigma(\cdot)$ denotes a non-linear activation function. Indeed, a prominent issue in the current mechanism of graph-based model is the lack of data dependency in the adjacency matrix A. As a result, even if the topology of the training sample changes, the existing model fails to fully utilize a proper and adaptive adjacency matrix to extract optimal features. This limitation can hinder the model's ability to effectively capture the underlying patterns and relationships in the data, leading to potential suboptimal performance in human parsing tasks. To address this issue, there is a need to devise approaches that can dynamically adapt the adjacency matrix based on the data characteristics, allowing model to leverage more appropriate and adaptive information during feature extraction and propagation. This motivates us to study the adaptive adjacency matrix (i.e., Graph Connection Adaption, GCA) to achieve a more effective propagation manner. Indeed, the adjacency matrix plays a crucial role in defining the topology of the graph by representing the connection relationships between the nodes. It encodes the edges or links between nodes, determining how they are interconnected in the graph structure. Consequently, the adjacency matrix governs the information propagation and neighborhood relationships in graph convolution layer, i.e., $A_{i,j} = \omega_{i,j} > 0$ if $e_{i,j} \in E$ and $A_{i,j} = 0$ if $e_{i,j} \notin E$. The left sketch in Fig. 6.1a presents the connection relationship of the fixed graph, where the edges formulate the inherent topological connection relationship. In Table 6.1, we show the accurate connection relationship of the fixed graph nodes for various datasets. Based on the analysis and the observation in Fig. 6.1b, when the right leg is occluded, the graph structure should be adaptively adjusted according to the content of the image. In this case, not only the graph node of the right leg needs to be discarded, but also the corresponding edges need to be deleted, as depicted by gray lines in Fig. 6.1b. The discarded nodes and deleted edges are depicted in gray lines to highlight their exclusion from the graph structure during the adaptive

Table 6.1 The connection relationship of the fixed graph nodes for various datasets

Dataset	Labels	Connection relationship
LIP (CIHP)	'1Hat', '2Hair', '3Glove', '4Sunglasses', '5Upper-clothes', '6Dress', '7Coat', '8Socks', '9Pants', '10Jumpsuits' ('Torso-skin'), '11Scarf', '12Skirt', '13Face', '14Left-arm', '15Right-arm', '16Left-leg', '17Right-leg', '18Left-shoe', '19Right-shoe'	1→[2, 13], 2→[1, 13], 3→[14, 15], 4→[13], 5→[6, 7, 9, 10, 11, 12, 14, 15], 6→[5, 7, 10, 11, 14, 15, 16, 17], 7→[5, 6, 9, 10, 11, 12, 14, 15], 8→[16, 17, 18, 19], 9→[5, 7, 10, 16, 17, 18, 19], 10→[5, 6, 7, 9, 11, 12, 13, 14, 15, 16, 17], 11→[5, 6, 7, 10, 13], 12→[5, 7, 10, 16, 17], 13→[1, 2, 4, 10, 11], 14→[3, 5, 6, 7, 10], 15→[3, 5, 6, 7, 10], 16→[6, 8, 9, 10, 12, 18], 17→[6, 8, 9, 10, 12, 19], 18→[8, 9, 16], 19→[8, 9, 17]
ATR	'1Hat', '2Hair', '3Sunglasses', '4Upper-clothes', '5Skirt', '6Pants', '7Dress', '8Belt', '9Left-shoe', '10Right-shoe', '11Face', '12Left-leg', '13Right-leg', '14Left-arm', '15Right-arm', '16Bag', '17Scarf'	1→[2, 11], 2→[1, 11], 3→[11], 4→[5, 6, 7, 14, 15, 17], 5→[4, 8, 12, 13], 6→[4, 8, 9, 10, 12, 13], 7→[4, 12, 13, 14, 15, 17], 8→[5, 6], 9→[6, 12], 10→[6, 13], 11→[1, 2, 3, 17], 12→[5, 6, 7, 9], 13→[5, 6, 7, 10], 14→[4, 7, 16], 15→[4, 7, 16], 16→[14, 15], 17→[4, 7, 11]
Pascal-Person-Part	'1Head', '2Torso', '3Upper Arms', '4Lower Arms', '5Upper Legs', '6Lower Legs'	1→[2], 2→[1, 3, 5], 3→[2, 4], 4→[3], 5→[2, 6], 6→[5]

adjustment process. To address this, we propose a strategy with adaptive matrix to achieve graph connection adaptation, as follows:

$$A_{adaptive} = A * Q, \tag{6.6}$$

where A is the fixed adjacency matrix which predefined based on the physical structure of the human body, $A_{adaptive}$ denotes the optimized adjacency matrix which can be adjusted adaptively according to training samples. The elements of matrix Q are solely parameterized through multilabel learning, implying that the optimized adjacency matrix $A_{adaptive}$ is entirely adjusted based on the training samples. Notably, the elements in Q tend to be close to either 0 or 1. By employing dot products in Eq. (6.6), GCA can effectively discard original connections, thereby enabling the model to adaptively adjust the graph structure based on the content of the input data. However, it is essential to mention that GCA is not capable of generating entirely new connections that are not present in the fixed graph. From the perspective of effective information propagation in graph reasoning, GCA enables model to propagate information more effectively, focusing on the most relevant and optimal features for each specific training sample. This adaptability in the graph structure allows model to extract and leverage crucial information, leading to enhanced performance and better representation learning in human parsing tasks. We change the Eq. (6.5) into the following form:

$$H_{l+1} = \sigma(\widetilde{D}_{adap}^{-\frac{1}{2}} \widetilde{A}_{adap} \widetilde{D}_{adap}^{-\frac{1}{2}} H_l W_l), \tag{6.7}$$

$$\widetilde{A}_{adap} = A_{adaptive} + I = A * Q + I, \tag{6.8}$$

where $\widetilde{D}_{adap}^{i,i} = \sum_j (\widetilde{A}_{adap}^{i,j})$. By combining GNA and GCA, the fixed graph can be optimized into a data-driven adaptive graph for each individual training sample. The resulting adaptive graph, which precisely captures the underlying data characteristics, is then employed to propagate information throughout the GCNs. This data-driven approach enhances the model's capacity to capture relevant and discriminative features, leading to improved performance and representation learning, ultimately resulting in more accurate and robust predictions across human parsing.

6.2.3 *Framework and Procedure*

In this chapter, the central theme of our approach is to model adaptive graphs and enhance the performance of human parsing tasks. To address this, we introduce a novel framework called Multilabel Learning based Adaptive Graph Convolutional Network (ML-AGCN). This framework is designed to dynamically generate adaptive graphs for various training samples by leveraging the principles of multilabel learning. Focusing on this, we use the graph node adaptation and graph connection adaptation to model this framework, as shown in Fig. 6.3.

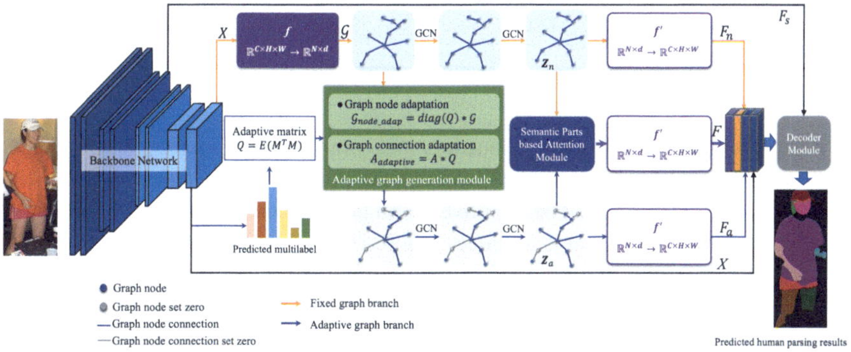

Fig. 6.3 Diagram of the proposed ML-AGCN

Fixed Graph Generation. In our network, we incorporate a nonlinear projection function that converts the deep spatial features into graph data. This function is depicted by the purple rectangular box in Fig. 6.3. Given the output features $X \in \mathbb{R}^{C\times H\times W}$ of the backbone, the first step involves reshaping it into a matrix $X \in \mathbb{R}^{C\times HW}$, effectively converting the spatial features into a vector representation. Subsequently, we design a nonlinear projection function that takes this vector representation X as input and converts it into the fixed graph data G, i.e., $\mathbb{R}^{C\times H\times W} \rightarrow \mathbb{R}^{N\times d}$, as follows:

$$G = f(X, W), \tag{6.9}$$

where W is the trainable transformation matrix, $G \in \mathbb{R}^{N\times d}$ is the fixed graph generated by the deep spatial features X, N is the number of the labeled semantic parts, d is the vector dimension of the node. This nonlinear projection function plays a crucial role in transforming the spatial features into a format suitable for constructing the fixed graph, enabling subsequent graph-based operations and feature propagation during the human parsing process. By performing this projection, we can effectively convert the spatial information of the input data into graph-based data, facilitating the graph reasoning and feature extraction processes in our approach. **Algorithm 1** clearly illustrates the process of Eq. (6.9).

Algorithm 1: The nonlinear projection function f

1: **Input**: The output features $X \in \mathbb{R}^{C\times H\times W}$ of the backbone, initialized trainable transformation matrix $W_1 \in \mathbb{R}^{C\times N}$, initialized trainable transformation matrix $W_2 \in \mathbb{R}^{C\times d}$
2: **Output**: The fixed graph data $G \in \mathbb{R}^{N\times d}$
3: Initialize the trainable transformation matrix W_1 and W_2
4: **for** each batch in training set **do**
5: Reshape $X \in \mathbb{R}^{C\times H\times W}$ to $X \in \mathbb{R}^{C\times HW}$
6: Compute $X^{'} = X^T W_1$
7: Compute $X^{''} = X^T W_2$
8: Compute $G = relu(X^{'T} X^{''})$
9: **end for**

Adaptive graph generation module. After generating the fixed graph, the next step involves designing the adaptive graph generation module, which is based on the GNA and GCA techniques. The adaptive graph generation module aims to create an adaptive graph structure that is uniquely tailored to each individual training sample. To achieve this, GNA is employed to facilitate adaptive graph node aggregation, allowing the model to dynamically adjust the graph node set based on the content of the input data. On the other hand, GCA enables the adaptive adjustment of graph connections, allowing the model to selectively retain or discard edges in response to the characteristics of each training sample. By integrating GNA and GCA, the adaptive graph generation module effectively constructs an adaptive graph for each training sample, enabling the model to make better use of the graph structure and optimize feature extraction and propagation. This adaptability enhances the model's performance and robustness in human parsing tasks, as it can adapt to variations and complexities in the input data, resulting in improved representation learning and more accurate predictions.

Semantic parts based attention module. As shown in Fig. 6.3, the ML-AGCN is an end-to-end network. Despite the limitations of the fixed graph, it is still able to capture rich structural features. Acknowledging the value of this information, the fixed graph is preserved within the framework to enrich the features. So, after the backbone, we construct a two-stream framework to learn two meaningful embeddings, i.e., optimized fixed graph embedding $\mathbf{Z}_n$ (depicted by orange lines in Fig. 6.3) and optimized adaptive graph embedding $\mathbf{Z}_a$ (depicted by blue lines in Fig. 6.3), respectively. Indeed, motivated by the goal of sufficiently fusing the rich information from the complex graph, we introduce a semantic parts based attention mechanism. This mechanism is designed to adaptively fuse the two meaningful embeddings, $\mathbf{Z}n$ and $\mathbf{Z}a$, to obtain the final human parsing result. The inputs to attentional layer are two meaningful embeddings $\mathbf{Z}_n = \left\{\overrightarrow{Z}_n^1, \overrightarrow{Z}_n^2, \ldots, \overrightarrow{Z}_n^N\right\}, \overrightarrow{Z}_n^i \in \mathbb{R}^{1\times d}$, and $\mathbf{Z}_a = \left\{\overrightarrow{Z}_a^1, \overrightarrow{Z}_a^2, \ldots, \overrightarrow{Z}_a^N\right\}, \overrightarrow{Z}_a^i \in \mathbb{R}^{1\times d}$, where d is the vector dimension of nodes, and N is the number of nodes. The attentional layer produces the final embedding $\widetilde{\mathbf{Z}} = \{\widetilde{Z}^1, \widetilde{Z}^2, \ldots, \widetilde{Z}^N,\}, \widetilde{Z}^i \in \mathbb{R}^{1\times d}$ with adaptively fusing $\mathbf{Z}_n$ and $\mathbf{Z}_a$:

$$\widetilde{\mathbf{Z}} = \boldsymbol{\alpha}_n * \mathbf{Z}_n + \boldsymbol{\alpha}_a * \mathbf{Z}_a. \tag{6.10}$$

To learn the weight matrices $\boldsymbol{\alpha}_n$ and $\boldsymbol{\alpha}_a$ for the semantic parts based attention mechanism, a trainable neural network is employed. This network acts as the attention mechanism and is capable of dynamically determining the importance or weight assigned to different parts of the embeddings $\mathbf{Z}_n$ and $\mathbf{Z}_a$. Let's focus on a specific node, denoted as $\overrightarrow{Z}_a^i \in \mathbb{R}^{1\times d}$, as an example to explain the procedure of the attention module. We firstly transform the embedding $\overrightarrow{Z}_a^i$ through a learnable nonlinear transformation (i.e., one-layer MLP), and then apply a semantic part based attention vector $e \in \mathbb{R}^{d'\times 1}$ to compute the importance of the node, denoted as ω_a^i:

$$\omega_a^i = e^T \cdot tanh(\psi \cdot (\overrightarrow{Z}_a^i)^T + b), \tag{6.11}$$

where $\psi \in \mathbb{R}^{d'\times d}$ is the trainable weight matrix, $b \in \mathbb{R}^{d'\times 1}$ is the bias vector. Note that for the embedding $\overrightarrow{Z}_n^i$, we adopt the same parameters to compute the semantic parts based attention value ω_n^i. After obtaining the attention values ω_a^i and ω_n^i, we normalize them with softmax function:

$$\alpha_a^i = \text{softmax}(\omega_a^i) = \frac{exp(\omega_a^i)}{exp(\omega_a^i) + exp(\omega_n^i)}, \tag{6.12}$$

$$\alpha_n^i = \text{softmax}(\omega_n^i) = \frac{exp(\omega_n^i)}{exp(\omega_a^i) + exp(\omega_n^i)}. \tag{6.13}$$

Obviously, the attention weights α_a^i and α_n^i for each node provide a measure of the importance of the corresponding embeddings $\overrightarrow{Z}_a^i$ and $\overrightarrow{Z}_n^i$ in the overall human parsing task. Larger values of α_a^i or α_n^i indicate higher importance of the corresponding embeddings. By computing the attention weights for all nodes, we can generate the weight matrices $\boldsymbol{\alpha}_n = diag([\alpha_n^i])$ and $\boldsymbol{\alpha}_a = diag([\alpha_a^i])$. The semantic parts based attention module then utilizes these attention weights to adaptively fuse the two meaningful embeddings $\mathbf{Z}_n$ and $\mathbf{Z}_a$ into a unified feature space. This fusion process optimizes the relationships between the two embeddings, allowing for a more comprehensive node embedding that captures the most relevant and discriminative information from both embeddings. In this way, the semantic parts based attention module enhances the feature representation by selectively focusing on the most informative parts of the embeddings, leading to improved performance in the human parsing task. This adaptive fusion mechanism further contributes to the overall end-to-end design of ML-AGCN, enabling it to achieve better representation learning and more accurate human parsing results.

Human parsing Loss. Once the embeddings $\mathbf{Z}_n$, $\mathbf{Z}_a$, and $\widetilde{\mathbf{Z}}$ are obtained from the graph convolutional layer, the subsequent step is to re-project this graph data back into spatial features F_n, F_a, and F using a nonlinear projection function $f^{'}$:

$$F_n, F_a, F = f^{'}(\mathbf{Z}_n, \mathbf{Z}_a, \widetilde{\mathbf{Z}}). \tag{6.14}$$

This re-projection process is essential to transform the graph-based embeddings back into the original spatial domain, preserving the spatial context and information of the features. By mapping the graph data to spatial features, the model can integrate both the graph-based structural knowledge and the spatial context of the input data. As we know, spatial information is crucial for achieving accurate and effective human parsing results. To leverage spatial information in the final experimental results, the approach utilizes the following steps:

1. Concatenation of Features: The output of the backbone network, denoted as X, is concatenated with the re-projected spatial features F_n, F_a, and F obtained from the graph convolutional layer. After a convolutional layer, this concatenation allows for a more comprehensive representation and enriched features.
2. Decoder Module: To further enhance the utilization of spatial information, a decoder module is designed. This module takes the low-level visual features F_s (e.g., from Res2 in ResNet-101) as spatial information complementary to the high-level semantic features. Concretely, the decoder module consists of a convolution layer followed by a bilinear interpolation upsampling layer. The two steps can be represent as following:

$$\widehat{y} = Decoder(conv(cat(F_n, F_a, F, X)), F_s). \tag{6.15}$$

By incorporating spatial information through the concatenation of features and utilizing the decoder module, the approach ensures that both high-level semantic features and low-level visual features are considered, leading to more comprehensive and accurate human parsing predictions. Given the predicted human parsing results $\widehat{y}$ and the corresponding ground truths y, we apply the cross-entropy loss as training objectives, as follows:

$$L_{parsing} = \frac{1}{pix} \sum_{pix} \sum_{N} -ylogp(\widehat{y}), \tag{6.16}$$

where pix represents the number of pixels in the image, N is the number of the labels.

Multilabel Learning Loss. In Sect. 6.2.1, the estimating multilabel classification score $M \in \mathbb{R}^{1\times N}$ is obtained, where N is the number of classes. The values of M range from 0 to 1, representing the probability or confidence of each class being present in the input image. Let's assume that the ground truth of the image is denoted as $\mathscr{M} \in \mathbb{R}^{1\times N}$, where $\mathscr{M}^t = \{0, 1\}$ indicates whether class t appears in the image or not. The value of 1 in $\mathscr{M}$ indicates the presence of the corresponding class, while 0 indicates its absence. To train the model for multilabel classification, the Binary Cross Entropy (BCE) loss is used as the objective. The BCE loss measures the difference between the predicted multilabel classification score M and the ground truth $\mathscr{M}$ for each class separately:

$$L_{multilabel} = -\frac{1}{N} \sum_{t=1}^{N} \mathscr{M}^t log(M^t) + (1 - \mathscr{M}^t) log(1 - M^t). \tag{6.17}$$

By minimizing the multilabel learning loss, the model can effectively learn to predict the presence or absence of multiple labels in the input image, and in turn, facilitate the human parsing task by associating semantic labels with corresponding regions of the image.

Label Consistency Loss. A label consistency loss is introduced at the label level to ensure consistency between the predicted multilabel and the predicted human parsing results. The main objective of the label consistency loss is to enforce the same multilabel predictions for both subtasks, explicitly aligning the multilabel predictions with the predicted human parsing results. Mathematically, the label consistency loss can be formulated as follows:

$$L_{label\_consistency} = -\frac{1}{N} \sum_{t=1}^{N} (\widetilde{M}^t - M^t)^2, \tag{6.18}$$

where $\widetilde{M}^t$ denotes the estimating multilabel classification score generated from the predicted human parsing results. The label consistency loss penalizes the differences between the predicted multilabel scores and the predicted human parsing results, aiming to minimize the discrepancies between the two sets of predictions. By incorporating the label consistency loss, ML-AGCN is explicitly designed to encourage and maintain the commonalities between multilabel learning and human parsing tasks. This alignment of objectives allows the framework to simultaneously optimize the accuracy of the adaptive graph generation and refine the human parsing results. In brief, the overall training objective is,

$$L = \lambda_1 L_{parsing} + \lambda_2 L_{multilabel} + \lambda_3 L_{label\_consistency}, \tag{6.19}$$

where λ_1, λ_2 and λ_3 are hyper-parameters to balance the contribution among $L_{parsing}$, $L_{multilabel}$, and $L_{label\_consistency}$. Finally, we train the whole framework in an end-to-end manner.

6.2.4 Experimental Results and Discussions

Compared with State-of-the-Arts. In our study, we present the quantitative results achieved by our proposed method and compare them with state-of-the-art approaches on the four challenging datasets, i.e., PASCAL-Person-Part [15], ATR [16], LIP [17], CIHP [18]. The results are summarized in Tables 6.2, 6.3, 6.4 and 6.5.

PASCAL-Person-Part [15]. In Table 6.2, we present the performance of our ML-AGCN approach and 11 state-of-the-art methods on the PASCAL-Person-Part test set. Our approach achieves an impressive mIoU score of 80.84%, surpassing all other state-of-the-art works by a significant margin. Notably, our method outperforms previous algorithms, such as Graphonomy [4], GPM [5], and PCNet [19],

Table 6.2 Comparison of human parsing performance with 11 state-of-the-arts on PASCAL-Person-Part test set [15]

Methods	Head	Torso	u-arm	l-arm	u-leg	l-leg	bkg	mIoU
MMAN (2018ECCV) [22]	82.58	62.83	48.49	47.37	42.80	40.40	94.92	59.91
WSHP (2018CVPR) [23]	87.15	72.28	57.07	56.21	52.43	50.36	**97.72**	67.60
PGN (2018ECCV) [18]	**90.89**	75.12	55.83	64.61	55.42	41.57	95.33	68.40
DeepLabv3+ (2018ECCV) [24]	–	–	–	–	–	–	–	68.60
JPPNet (2019TPAMI) [17]	83.26	62.40	47.80	45.58	42.32	39.48	94.68	59.36
CNIF (2019ICCV) [20]	88.02	72.91	64.31	63.52	55.61	54.96	96.02	70.76
Graphonomy (2020TPAMI) [4]	–	–	–	–	–	–	–	71.14
SCHP(2020TPAMI) [25]	87.41	73.80	64.98	64.70	57.43	55.62	96.26	71.46
GPM (2020AAAI) [5]	–	–	–	–	–	–	–	71.65
PRHP (2021TPAMI) [21]	89.73	75.22	66.87	66.21	58.69	58.17	96.94	73.12
PCNet (2020CVPR) [19]	90.04	76.89	69.11	68.4	60.78	60.14	96.78	74.59
Ours	90.62	**82.17**	**74.99**	**74.05**	**74.36**	**72.25**	97.46	**80.84**

Table 6.3 Comparison of human parsing performance with 12 state-of-the-arts on the LIP validation set [17]. The symbol † marks the multiple-scales testing results

Methods	Hat	Hair	Glove	s-glass	u-clot	Dress	Coat	Sock	Pant	j-suit	Scarf
DeepLab† [28]	59.76	66.22	28.76	23.91	64.95	33.68	52.86	37.67	68.05	26.15	17.44
SSL† [26]	58.21	67.17	31.20	23.65	63.66	28.31	52.35	39.58	69.40	28.61	13.70
MMAN [22]	57.66	65.63	30.07	20.02	64.15	28.39	51.98	41.46	71.03	23.61	9.65
JPPNet† [17]	63.55	70.20	36.16	23.48	68.15	31.42	55.65	44.56	72.19	28.39	18.76
DeepLabv3+ [24]	65.02	71.06	37.98	31.37	68.33	28.25	54.14	47.67	73.99	27.12	17.92
CE2P [29]	**87.67**	65.29	**72.54**	39.09	32.73	**69.46**	32.52	**56.28**	49.67	74.11	27.23
CorrPM [27]	–	–	–	–	–	–	–	–	–	–	–
PCNet [19]	69.32	73.08	44.72	34.21	**72.59**	36.02	**60.84**	51.03	76.66	**38.78**	31.60
OCR [30]	67.42	73.16	44.01	36.31	69.90	35.53	57.03	51.87	75.09	29.85	19.16
SLRS [31]	–	–	–	–	–	–	–	–	–	–	–
CNIF† [20]	69.55	73.45	45.17	**41.45**	70.57	38.52	57.94	54.02	75.07	28.00	**31.92**
SCHP [25]	69.96	73.55	50.46	40.72	69.93	39.02	57.45	54.27	76.01	32.88	26.29
SCHP† [25]	70.63	**74.09**	51.40	41.70	70.56	40.06	58.17	55.17	**76.57**	33.78	26.63
Ours	69.94	73.31	49.72	41.00	70.63	36.91	58.50	53.68	76.05	32.03	27.54
Methods	Skirt	Face	l-arm	r-arm	l-leg	r-leg	l-shoe	r-shoe	bkg	mIoU	
DeepLab† [28]	25.23	70.00	50.42	53.89	39.36	38.27	26.95	28.36	84.09	44.80	
SSL† [26]	22.52	74.84	52.83	55.67	48.22	47.49	31.80	29.97	84.64	46.19	
MMAN [22]	23.20	69.54	55.30	58.13	51.90	52.17	38.58	39.05	84.75	46.81	
JPPNet† [17]	25.14	73.36	61.97	63.88	58.21	57.99	44.02	44.09	86.26	51.37	
DeepLabv3+ [24]	25.00	73.99	63.64	65.69	57.35	56.99	43.97	44.74	87.51	52.09	
CE2P [29]	14.19	22.51	**75.50**	65.14	66.59	60.10	**58.59**	46.63	46.12	53.10	
CorrPM [27]	–	–	–	–	–	–	–	–	–	55.33	
PCNet [19]	**33.94**	**76.65**	67.07	68.74	60.22	60.16	47.65	48.67	**88.68**	57.03	
OCR [30]	26.63	75.95	66.81	68.65	63.23	62.12	50.60	50.68	87.96	55.60	
SLRS [31]	–	–	–	–	–	–	–	–	–	56.34	
CNIF† [20]	30.20	76.38	68.28	69.49	65.52	65.51	52.67	53.38	87.99	57.74	
SCHP [25]	31.68	76.19	68.65	**70.92**	67.28	66.56	55.76	56.50	88.36	58.62	
SCHP† [25]	32.83	76.63	69.33	71.76	**67.93**	**67.42**	56.56	**57.55**	88.40	**59.36**	
Ours	31.35	75.96	68.56	71.24	67.48	66.94	55.05	56.35	88.28	58.53	

Table 6.4 Comparison of pixel accuracy and F1-score on ATR test [16]

Methods	Pixel accuracy	F1-score
ATR (2015TPAMI) [16]	91.11	64.38
PSPNet (2017CVPR) [14]	95.20	75.84
Deeplabv3+ (2018ECCV) [24]	95.96	79.49
TGPNet (2018ACMM) [32]	96.45	81.76
CNIF (2019ICCV) [20]	96.26	85.51
SCHP (2020TPAMI) [25]	96.25	85.55
Graphonomy (Universal)(2020TPAMI) [4]	98.03	**89.24**
PRHP (2021TPAMI) [21]	96.84	87.25
Ours	97.52	88.91

Table 6.5 Comparison of multiple human parsing performance on the CIHP validation set [18]

Methods	Mean accuracy	mIoU
PGN (2018ECCV) [18]	64.22	55.80
DeepLabv3+ (2018ECCV) [24]	67.69	58.95
CE2P (2019AAAI) [29]	–	59.50
Graphonomy (Universal) (2020TPAMI) [4]	65.75	57.78
Graphonomy (PASCAL) (2020TPAMI) [4]	66.65	58.58
GPM (2020AAAI) [5]	68.95	**60.36**
Ours	71.37	59.56

which achieve mIoU scores of 71.14, 71.65, and 74.59%, respectively. However, all of these methods predefine a fixed graph based on the human prior structure for all the training samples in the training process. Indeed, the fixed graph, which remains static and uniform across all training samples, is not ideal for achieving better representation learning, especially in the context of human parsing tasks. The inherent limitation of a fixed graph lies in its inability to adaptively integrate node features, leading to suboptimal performance when dealing with diverse and complex human data. Our proposed ML-AGCN approach addresses this limitation by introducing a data-driven graph generation mechanism. This data-driven approach allows the model to dynamically adjust the node sets and edge sets of the graph for each individual training sample. By doing so, the adaptive graph can effectively capture the specific structural information present in each image, leading to more accurate and effective feature learning. The ability of our approach to generate a data-driven graph is a key strength that allows it to flexibly integrate node information based on the content of each image. As a result, ML-AGCN can better represent the complex and diverse relationships between human body parts, leading to more precise and reliable human parsing results. In the PASCAL-Person-Part dataset, the challenge of having fewer training images has motivated some competing approaches to leverage the hierarchical structure of the human body to enhance the performance of human parsing. Approaches like Contextual Nested Inference Framework (CNIF) [20] and Progressively Refined Hierarchical Parsing (PRHP) [21] utilize multi-level graphs to capture correlation information among multi-granularity annotations, aiming for more efficient and complete human parsing. While these methods employ mutual learning across multi-granularity annotations, our ML-AGCN approach takes a different approach by focusing on generating adaptive graphs and distilling high-level graph representations through the optimal fusion of fixed and adaptive graphs. This adaptivity gives ML-AGCN a significant advantage over methods like CNIF and PRHP, which rely on predefined multi-level graphs that may not be as effectively tailored to the characteristics of individual training samples. As we can see in Table 6.2, ML-AGCN achieves superior performance, surpassing CNIF by 10.08% and PRHP by 7.72%. This performance gain is a testament to the power of adaptive graph generation and its ability to enhance the representation learning process, ultimately leading to more accurate and comprehensive human parsing results.

LIP [17]. The LIP is the largest single human parsing dataset, including a large number of training images and annotations for 19 types of labels. Due to these benefits, LIP has become the gold standard benchmark for human parsing tasks. In Table 6.3, we present a performance comparison of our ML-AGCN approach with 12 previous methods on the LIP validation set. To provide a detailed view of the performance of these algorithms, we report the per-class IoU in Table 6.3. Among the previous methods, SCHP [25] achieves the best results with a mIoU of 58.62%. Additionally, when employing data augmentation techniques for all the validation data, SCHP's mIoU is further improved to 59.36%. It is worth noting that obtaining accurate pixel-level annotations in human parsing can be challenging, especially for datasets like LIP with fine granularity annotations, may contain some incorrect labels. SCHP applies a cyclically learning scheduler to optimize those incorrect masks and then utilizes the corrected mask to further re-train the network. In this online correction, SCHP effectively addresses the interference caused by these label noises and achieves the best performance. Despite the label noises introduced by fine-grained annotations, our ML-AGCN approach demonstrates its effectiveness and robustness on the LIP dataset. Indeed, our ML-AGCN approach leverages the special structure prior information of the human body to explore high-level features and generate adaptive graphs through a data-driven approach, leading to a performance boost in the framework. Notably, even without the online correction, our approach achieves competitive results with a mIoU of 58.53%, compared to SCHP's 58.62%. This demonstrates the efficacy of our data-driven adaptive graph generation in accurately segmenting human body parts. Some other methods in the literature also utilize structural knowledge of human pose to facilitate human parsing, such as Semantic Structure Learning (SSL) [26], Joint Human Parsing and Pose Estimation Network (JPPNet) [17], and Contextual Correlation Prior for Multi-Human Parsing (CorrPM) [27]. However, these methods require additional inputs and annotations, such as human pose and key points, which can increase the computational complexity and introduce potential errors. In contrast, our ML-AGCN approach does not rely on additional inputs or annotations, making it more efficient and accessible for practical applications. Furthermore, our approach achieves significant performance improvements over these methods. Specifically, ML-AGCN surpasses SSL by 12.34%, JPPNet by 7.16%, and CorrPM by 3.2% in terms of mIoU. The superior performance of ML-AGCN compared to these pose-based methods highlights the strength of our data-driven adaptive graph generation and fusion strategy. By fully utilizing the structural information inherent in the human body, ML-AGCN can effectively capture complex relationships between body parts and generate more accurate and comprehensive human parsing results.

ATR [16]. Similar to LIP, ATR also serves as a fine-grained benchmark, consisting of 18 semantic categories. ATR is specifically designed for fashion AI, where the "person" category plays a significant role in the images. The results of ATR are presented in Table 6.4. Notably, Graphonomy(Universal) represents a universal human parsing model that operates across multiple domains. Our method exhibits impressive performance, achieving a high F1-score of 88.91%, and surpassing most previous approaches by a considerable margin. However, Graphonomy(Universal), which constructs a fixed graph for each dataset and leverages graph transfer learning across domains, outperforms our proposed method by 0.51% in pixel accuracy and 0.33% in F1-score. The utilization of graph transfer learning in Graphonomy(Universal) enables the capture of global semantic coherence, leading to improved human parsing performance. While our proposed method may perform slightly worse than Graphonomy(Universal) on the ATR dataset, it demonstrates superior performance on the PASCAL-Person-Part dataset and CIHP dataset. Furthermore, when compared to SCHP [25], our method slightly lags behind on the LIP dataset. However, it achieves substantial improvements of 1.27% and 3.36% in pixel accuracy and F1-score, respectively, on the ATR dataset. Overall, the reported quantitative results provide strong evidence for the effectiveness of our proposed method on multiple datasets.

CIHP [18]. CIHP stands as the most popular benchmark for multiple human parsing, providing annotations for 19 distinct labels. To demonstrate robustness in multiple human parsing, we conducted further testing of ML-AGCN on the CIHP dataset and compared it against several representative approaches, as depicted in Table 6.5. Notably, Graphonomy (PASCAL) refers to a method that transfers semantic features distilled from the PASCAL-Person-Part dataset to CIHP, thereby enriching the graph features. The results in Table 6.5 demonstrate that our proposed method outperforms the majority of state-of-the-art approaches in terms of mIOU. Particularly, when compared to Graphonomy [4], which utilizes a fixed graph for human parsing and performs graph transfer learning across different datasets, our method achieves a superior mIOU by 0.98% over Graphonomy (Universal) and 1.78% over Graphonomy (PASCAL). The reason is mainly that CIHP annotates 19 types of labels, so the optimal graph for each training sample varies in both size and connectivity. In this scenario, the adaptive graph generation of ML-AGCN effectively addresses this challenge. Moreover, ML-AGCN comprehensive utilizes of fixed, adaptive, and fusion graphs which further enhances performance. When compared to GPM, our method exhibits only a slight reduction in mIOU. However, it achieves considerable gains in mean accuracy, reaching 71.37%. Additionally, our approach significantly surpasses GPM by 9.19% on the PASCAL-Person-Part dataset.

Qualitative Results of Four Types of Hard Samples. There are four types of hard samples in the human parsing task, namely: (i) absence of limbs, (ii) complex body postures, (iii) severe occlusions, and (iv) dense crowd scenarios. The visual experimental results for these challenging samples are presented in Figs. 6.4, 6.5, 6.6, 6.7, and 6.8. Figure 6.4a illustrates instances where human bodies exhibit severe limb absence, resulting in substantial parsing errors characterized by indistinct boundaries and erroneous semantic label assignments. However, leveraging the comprehensive node embedding derived from the adaptive graph convolutional network, our proposed approach effectively mitigates the adverse effects of severe limb absence, yielding precise parsing results that align closely with ground truth annotations. In Fig. 6.5b, the hard samples with complex body postures and viewpoints

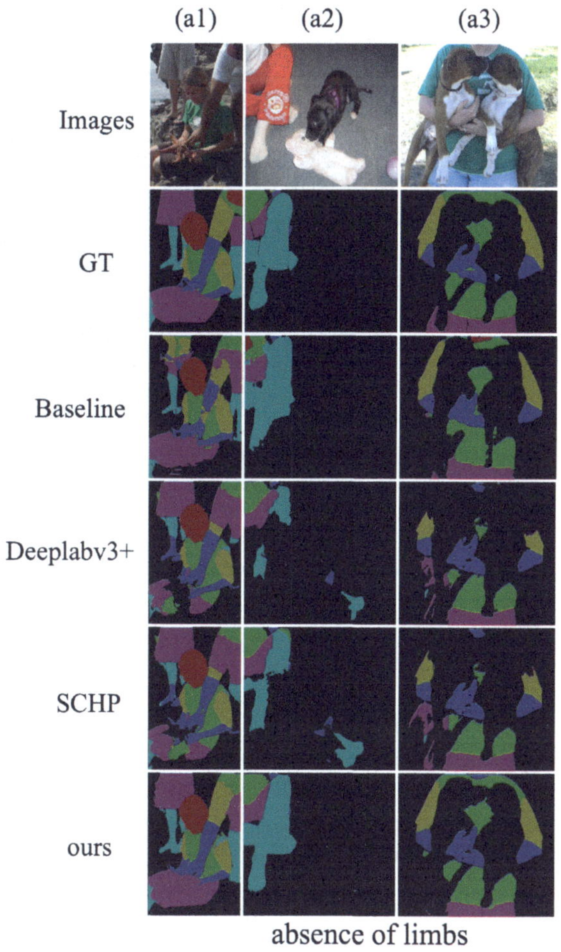

Fig. 6.4 Visualized comparison of human parsing results on PASCAL-Person-Part test set

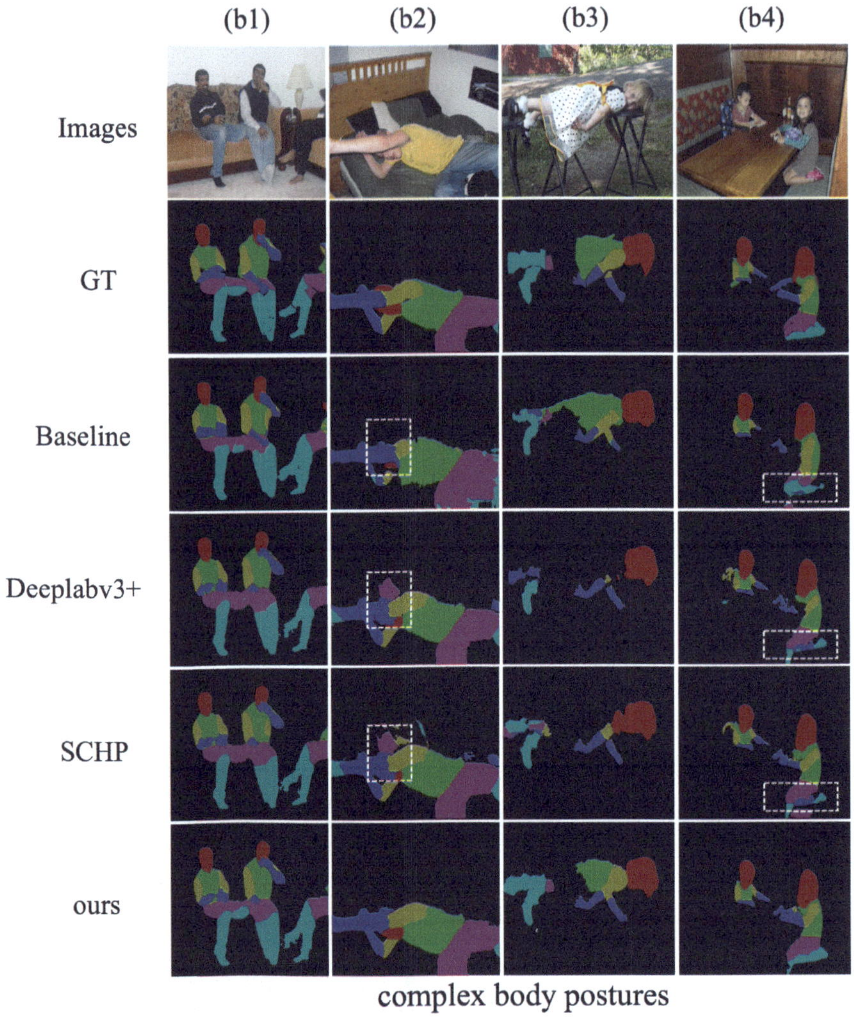

Fig. 6.5 Visualized comparison of human parsing results on PASCAL-Person-Part test set

are still a challenging task for the existing algorithms. As depicted in Fig. 6.5b2, even for the difficult scenario of a rare pose, most algorithms exhibit notable deficiencies in accurately parsing the head region. Leveraging the advantages of an adaptive graph, our proposed approach constructs an optimal graph tailored to handle such challenging samples, enabling effective graph reasoning and extraction of high-level features. Consequently, the proposed method excels in processing these samples, yielding precise parsing results, as evidenced in the comparable parsing results depicted in Fig. 6.5b4. Another type of hard sample is shown in Fig. 6.6c, there are severe occlusion problems caused by the external environment or other

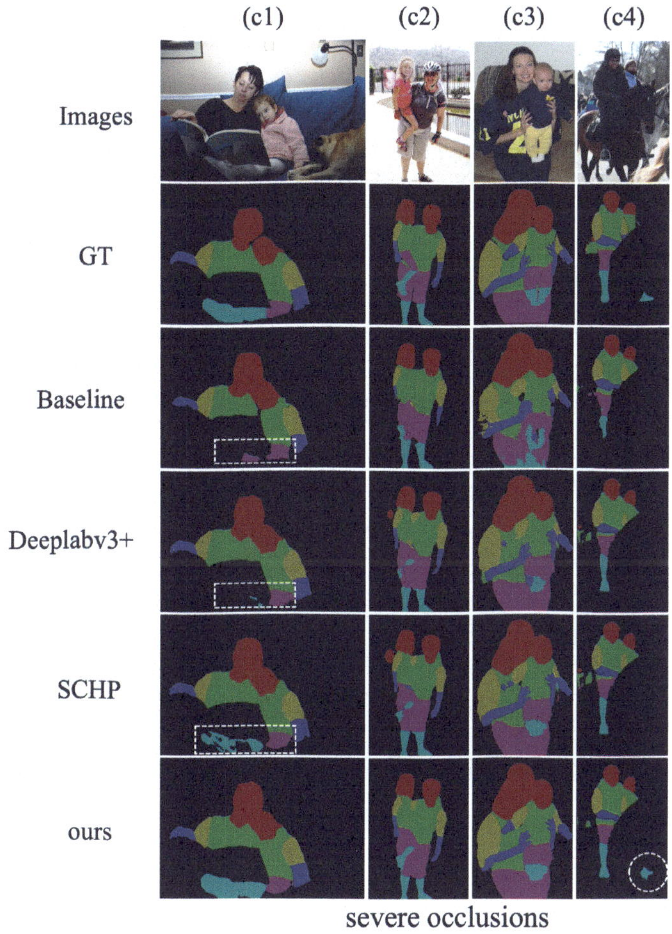

Fig. 6.6 Visualized comparison of human parsing results on PASCAL-Person-Part test set

human bodies in the images. This undoubtedly harms the training processing and decreases the accuracy of predictions. As depicted in Fig. 6.6c1, we can observe that there are severe inconsistencies in the predictions on the lower legs, as outlined by the white dashed rectangle. However, leveraging the advantages of our semantic parts based attention module, which optimizes the relationships between the embeddings to obtain a comprehensive node embedding, our proposed approach achieves a highly accurate parsing result. Moreover, our approach demonstrates superior performance in accurately parsing the left foot, as evidenced in Fig. 6.6c4, depicted by the white dashed circle. In this particular instance, a significant portion of the left leg is obscured, leaving only the left foot visible. Adding to the complexity, the color of the left foot is similar to the background color. Consequently, existing comparative

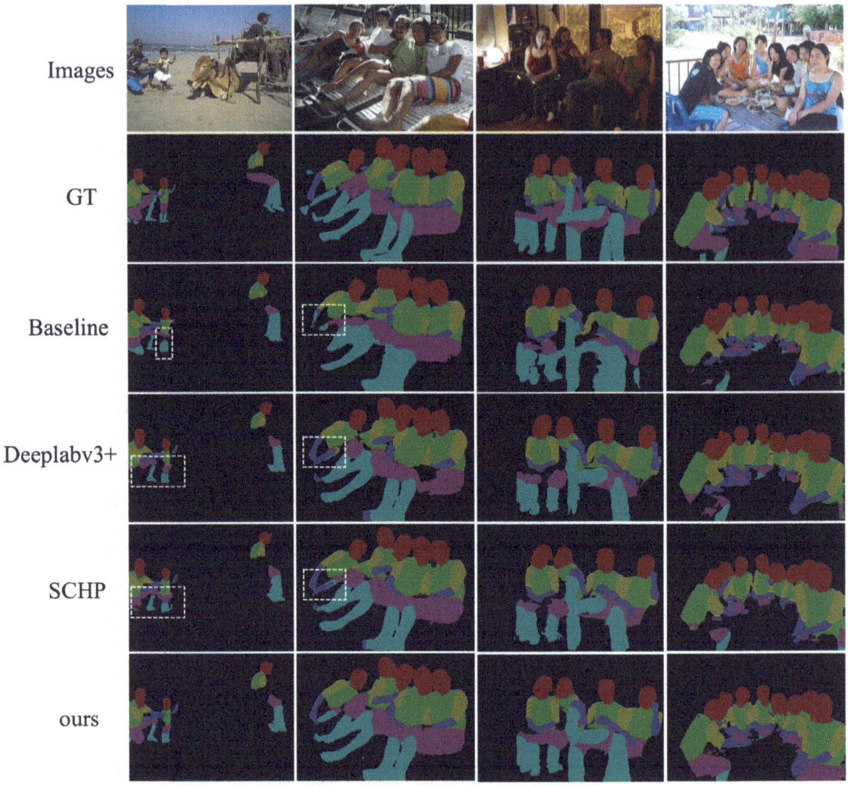

Fig. 6.7 Visualized comparison of dense crowd results on PASCAL-Person-Part test set

approaches struggle to accurately parse this region; however, our approach successfully addresses this challenge. Furthermore, in addition to the previously discussed challenges, handling a dense crowd of multiple individuals is a difficult challenge for accurate human parsing, as depicted in Figs. 6.7 and 6.8. These images exhibit numerous human bodies with complex mutual occlusions. Additionally, the presence of various illumination changes further exacerbates the difficulty in correctly parsing the human body. Nonetheless, owing to the efficient graph reasoning facilitated by our proposed framework, we observe that our approach surpasses other comparative methods, yielding precise parsing results characterized by clear boundaries despite these challenging conditions. It is worth noting that our approach consistently demonstrates the ability to accurately parse intricate details of human bodies even within chaotic crowd scenes and complex backgrounds, as highlighted within the white dashed rectangle. These compelling visual results collectively affirm the effectiveness and robustness of our proposed method.

Visualization of the Learned Adaptive Graphs. To provide a more intuitive demonstration of the distinction between a fixed graph and the learned adaptive

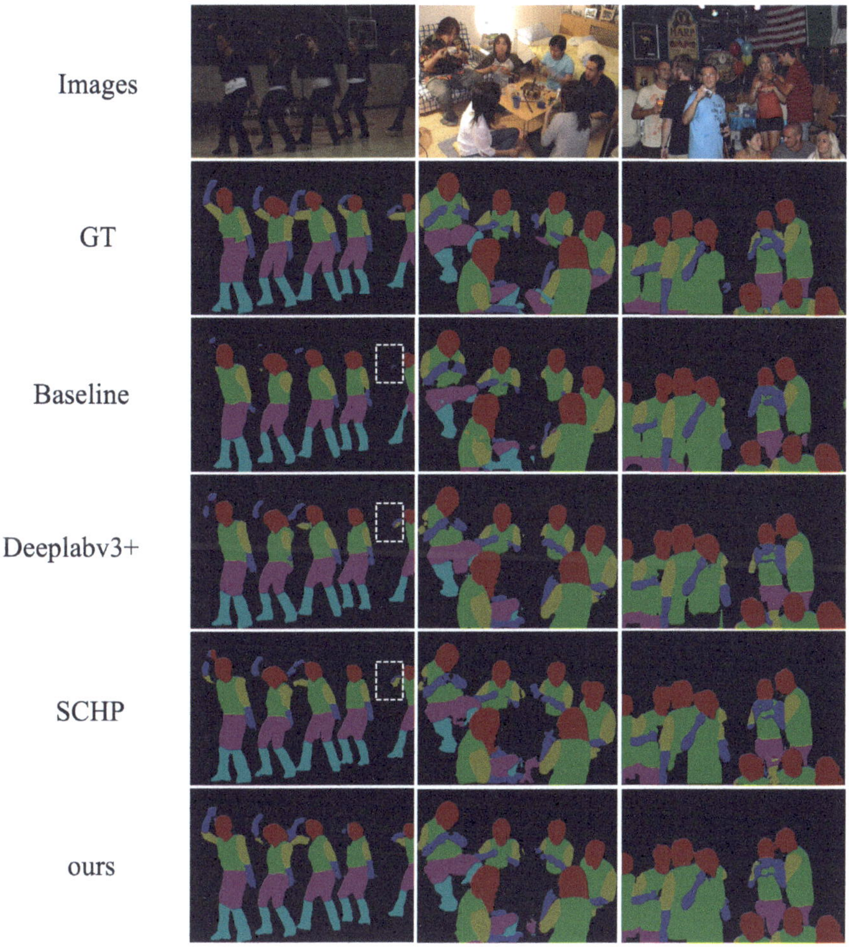

Fig. 6.8 Visualized comparison of dense crowd results on PASCAL-Person-Part test set

graph, we present visualizations of the corresponding adjacency matrices derived from the LIP validation set. These visualizations are illustrated in Fig. 6.9.

As previously mentioned, the LIP dataset defines a total of 19 semantic human part labels. Consequently, the size of the adjacency matrix is 19×19. The values within the adjacency matrix indicate the strength of connections and are visually represented by blue blocks with varying intensities, as depicted in Fig. 6.9. It is evident that the fixed adjacency matrix significantly differs from the original adjacency matrix. In contrast, our proposed approach generates an adaptive adjacency matrix that closely aligns with the original adjacency matrix, as demonstrated in the visualizations.

Effect of Adaptive Graph Generation Module. We proceed to provide a comprehensive demonstration of how the adaptive graph generation module significantly

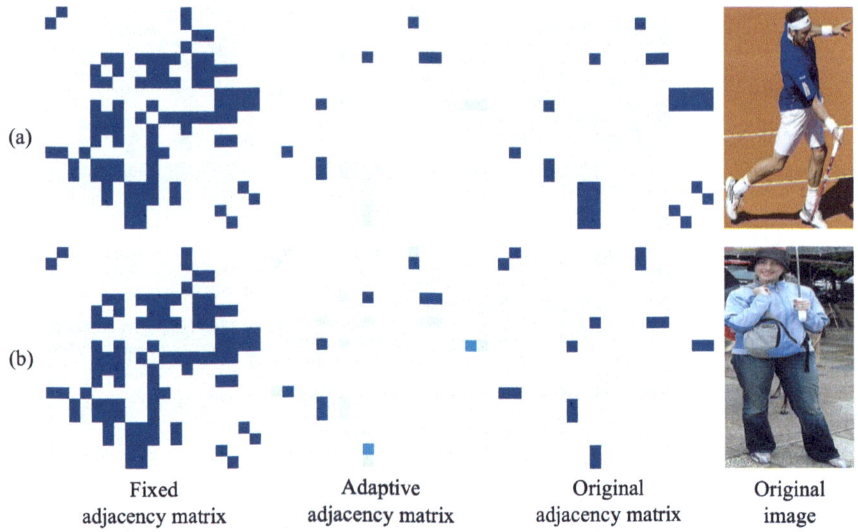

Fig. 6.9 Visualized of the adjacency matrix

Table 6.6 The effect of adaptive graph generation module is evaluated on LIP validation set

#	Fixed graph	GNA	GCA	Pixel accuracy	Mean accuracy	mIoU
1	×	×	×	87.62	70.00	55.86
2	√	×	×	88.04	70.30	56.37
3	√	√	×	88.06	70.69	58.22
4	√	×	√	87.86	70.45	57.88
5	√	√	√	88.25	71.24	**58.53**

contributes to our framework's efficacy and further conduct a detailed analysis of the underlying reasons for this effectiveness. The summarized results of these analyses are presented in Table 6.6. Specifically, we assess the performance of the baseline model, which comprises ResNet-101 and the context encoding module (PSP), denoted as #1. To assess the impact of the adaptive graph generation module, we compare the proposed method with and without the adaptive graph generation module (#2 vs #5). Notably, we achieve the optimal performance, attaining a mIoU of 58.53%, through the simultaneous utilization of both the fixed graph and the adaptive graph. This integration yields a substantial improvement of 2.67% over the baseline method in terms of mIoU. Unsurprisingly, the comparative analysis between #2 and #5 underscores that relying solely on the fixed graph is insufficient for optimal representation learning and fails to adequately capture the data diversity inherent in the human parsing task. Furthermore, we conduct separate evaluations to ascertain the individual contributions of GNA and GCA within the adaptive graph generation module, as illustrated in #3 and #4. Through the exclusive utilization of GNA, we observe a notable enhancement in performance, elevating mIoU from 55.86% to 58.22%. Conversely, employing GCA yields a slightly lower improvement, resulting in a mIoU of 57.88%. The comparison (#3 versus #4) reveals that GNA makes a more significant contribution compared to GCA. This can be attributed to GNA's effective pooling strategy, which dynamically filters features from nodes, effectively removing potential noise. Consequently, each node can adaptively propagate with distinct subsets of neighbors across different training samples. Owing to this advantage, the GNA, which can help the network to learn more abundant and accurate features, tends to perform better. Remarkably, it is evident that both GNA and GCA components mutually reinforce each other, leading to superior performance throughout the training process. To facilitate a more intuitive understanding of the distinct contributions of these components, we provide visual results from the ablation experiments conducted on the adaptive graph generation module, as depicted in Fig. 6.10. In Fig. 6.10a, the presence of label noises (highlighted by white dotted rectangles)

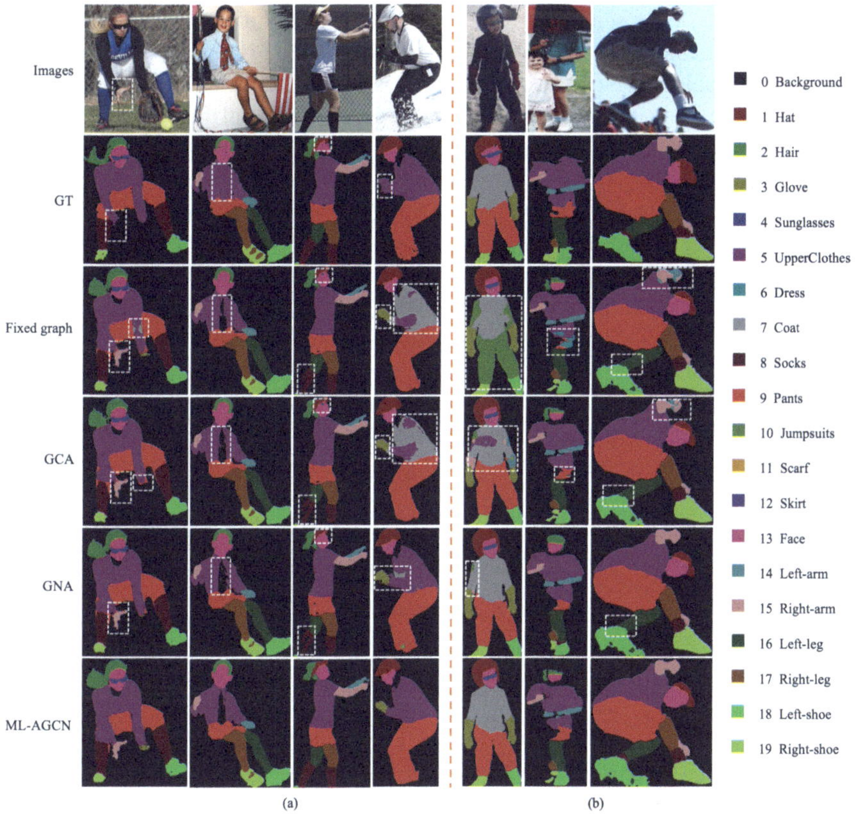

Fig. 6.10 Visualization results of the ablation experiments for adaptive graph generation module on the LIP validation set

inevitably introduces inaccuracies in feature extraction, thereby undermining the overall framework's performance. Notably, it becomes apparent that the mere addition of GCA or GNA does not suffice to yield significantly improved parsing results. The proposed method, which integrates GCA with GNA, effectively mitigates the adverse effects of label noises and consistently achieves superior performance. In Fig. 6.10b, we present results for challenging samples characterized by complicated textures (column 1), severe occlusions (column 2), and complex body postures (column 3). Notably, addressing these challenging samples solely through the fixed graph proves to be inadequate. Rows 4 and 5 illustrate the improvements attained by incorporating either GCA or GNA. However, it is evident that the predictions of the proposed method consistently exhibit higher precision compared to individual additions of GCA or GNA.

Effect of Semantic Parts based Attention Module. In the presented framework, optimal experimental results are attained by concatenating all output features, comprising the backbone output denoted as X, the output of the fixed graph branch represented as F_n, the output of the adaptive graph branch denoted as F_a, and the output of the semantic parts-based attention branch represented as F. To assess the significance of the semantic parts based attention within the proposed method, we conduct a dedicated ablation study. The summarized results of this study are presented in Table 6.7. Notably, in this ablation study, we include X by default for comprehensive evaluation. The results for both the baseline and the proposed method are summarized in #1 and #6 respectively. In the initial phase of the study, we present the performance of the framework with individual application of F_n, F_a, or F, as reflected in #2, #3, and #4, respectively. Notably, the quantitative results from #2 and #3 exhibit a significant decrease compared to #4. This observation affirms that utilizing either the fixed graph or the adaptive graph in isolation is inadequate for achieving optimal performance in human parsing. Conversely, as indicated in #4, the application of the semantic parts-based attention module yields notably improved performance. This observation strongly suggests that the semantic parts-based attention module effectively distills a high-level graph representation by weighted fusion of features from the fixed graph and the adaptive graph. To further ascertain the compensatory role of the adaptive graph in mitigating the limitations of the fixed graph,

Table 6.7 The effect of semantic parts based attention module is evaluated on LIP validation set

#	F_n	F_a	F	Pixel accuracy	Mean accuracy	mIoU
1	×	×	×	87.62	70.00	55.86
2	√	×	×	88.04	70.30	56.37
3	×	√	×	87.83	72.07	57.82
4	×	×	√	88.11	71.27	58.32
5	√	√	×	88.10	71.13	58.31
6	√	√	√	88.25	71.24	**58.53**

we conduct an experiment by removing the semantic parts-based attention module from the configuration as depicted in Fig. 6.3. The resulting performance is captured in #5. As anticipated, #5 yields a mIoU of 58.31%, closely resembling the performance of #4. However, both #4 and #5 exhibit inferior performance when compared to the complete proposed framework (#6), where an mIoU of 58.53% is achieved. Consequently, our proposed method attains the optimal results by synergistically incorporating feature maps from the fixed graph branch, adaptive graph branch, and the semantic parts-based attention branch. In summary, the ablation experiments underscore the critical role of the semantic parts-based attention module in distilling a complex graph infused with rich information, thereby significantly enhancing the performance of human parsing.

6.3 Dynamic Graph Pooling for Human Parsing

In this chapter, we propose a dynamic graph pooling for human parsing. It enables the model to dynamically adjust the receptive fields during the pooling process, which is particularly crucial for capturing both local and global context in an image. With the dynamic graph pooling, the model can learn rich hierarchical feature by adaptively selecting and downsampling nodes or edges based on their importance. In this way, we can capture accurate and detailed features from fine-grained graph to coarse-grained graph. Dynamic graph pooling allows for the adaptive aggregation of information from the input graph, enabling the model to focus on important nodes or features relevant to human parsing. For the flexibility of the dynamic graph, the model can effectively handle the complex human poses, making it possible to obtain a high performance human parsing model.

6.3.1 Construction of Hierarchical Semantic Graph for Human Parsing

Definition of static hierarchical semantic graph. Traditional human parsing methods generally construct a semantic graph based on the structured nature of the human body and often overlook the highly structured hierarchy. For instance, the label head can be subdivided into hat, face and hair, the torso encompasses scarf, upper-clothes, and coat. Note that both the structured nature and the highly structured hierarchy can be considered as powerful semantic structural prior knowledge to boost the model learning in weakly supervised human parsing. In order to explain the highly structured hierarchy of the human body more clearly, we construct a static hierarchical semantic graph, as shown in Fig. 6.11. There are four semantic graphs in the static hierarchical semantic graph, i.e., G_0, G_1 , G_2, G_3. The dotted lines $\dashrightarrow$, $\dashrightarrow$, $\dashrightarrow$ represent the hierarchical structure-property between static semantic

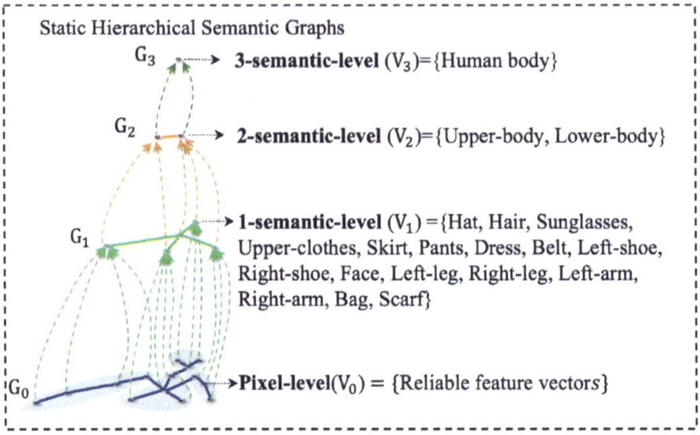

Fig. 6.11 Construction of hierarchical semantic graph for human parsing

graphs. Concretely, the static hierarchical semantic graph is grouped into four levels static semantic graphs, i.e., $G = \{G_0, G_1, G_2, G_3|G_n = (V_n, E_n, A_n)\}$, where V_n and E_n stand for the set of nodes and edges of static semantic graph G_n, A_n is the corresponding static adjacency matrix. Specially for dense graph G_0, we define the reliable pixel as the graph nodes and the corresponding adjacency matrix A_0 is computed by the pseudo labels. For the other three semantic graphs G_1, G_2, G_3, the graph node represent the labels of different granularity. For instance, the 1-semantic-level G_1 is the most fine-grained labels, involving the human parts and the fashion clothes, the 2-semantic-level G_2 only consists of the upper-body and lower-body, and the 3-semantic-level G_3 encompasses full-body and background. Noted that all the adjacency matrix A_1, A_2, A_3 strictly consistent with the natural topological connection of the human body, such as $A_1^{\text{hair,hat}} = 1$, $A_1^{\text{belt,pants}} = 1$, and $A_1^{\text{hair,pants}} = 0$. Here, we attend to take full advantage of the hierarchical structure representation for high performance human parsing model.

6.3.2 *Dynamic Topology-Aware Graph Pooling for Human Parsing*

Based on the hierarchical structure of human body, Wang et al. [21] present a novel approach for human parsing, focusing on leveraging the highly structured hierarchy of the human body to enhance performance. They analyze the effectiveness of this hierarchical representation, considering that the human body naturally encodes a hierarchical structure. Indeed, Graph Neural Networks (GNNs) have emerged as a successful extension of Convolutional Neural Networks (CNNs), allowing them to operate on non-Euclidean domains, such as graph datas. GNNs have shown significant promise in various applications. However, GNNs face challenges in distilling multiscale and hierarchical representations similar to the pooling operations in traditional CNNs. In CNNs, pooling operations, such as max-pooling or average-pooling, help downsample the feature maps, reducing their spatial dimensions while retaining important features. This pooling process allows CNNs to capture hierarchical and multiscale representations, which are crucial for tasks like object recognition and feature extraction. For graph pooling, various attempts have been made to address this issue. These methods involve clustering-based methods, graph coarsening strategies, or attention mechanisms. They aim to identify important nodes and relations within the graph. In this chapter, we propose a new graph pooling for human parsing based on the hierarchical semantic prior knowledge. By effectively utilizing this hierarchical semantic prior knowledge, we believe that it is possible to develop an improved weakly supervised model to achieve high-quality human parsing results. Here, we effectively encode the hierarchical semantic structural prior knowledge into the graph representation for human parsing. By generating the dynamic hierarchical semantic graph in a dense-to-coarse manner, we naturally captures and utilizes the structural information. Unlike conventional methods that rely on static semantic graphs for information propagation, Dynamic Topology-aware Pooling (DTP) is a novel pooling strategy. By collaborating with multilabel classification techniques, it enables information flow within dynamic semantic graphs. This dynamic nature allows for better utilization of the hierarchical semantic structural prior knowledge. DTP generates dynamic graphs and incorporate the hierarchical information which facilitates more effective pooling compared to traditional pooling methods. This, in turn, leads to improved results in weakly supervised human parsing. By dynamically adapting the graph representation from dense to coarse, the DTP can fully leverage the inherent structure of the human body, resulting in enhanced parsing performance.

6.3.3 Framework and Procedure

As shown in Fig. 6.12, we formulate the extraction of hierarchical structure representation as a graph pooling procedure, based on the insight that the dense graph is gradually coarsened into sparse semantic graphs until the full human body. Previous methods have verified the effectiveness of graph neural networks in human parsing. However, the graph neural networks only can aggregate and propagate the features in a flatten graph, and it fails to capture the hierarchical representations of the human body. In this case, on the basis of the human body understanding, we design a graph pooling named dynamic topology-aware pooling (DTP) to learn the hierarchical features for human parsing. DTP attends to encode the semantic structural prior knowledge into graph pooling operation to capture hierarchical structure representation for a better result. As shown in Fig. 5.10, the dense graph G_0 is gradually coarsened into sparse semantic graph G_1, G_2, G_3 through the proposed DTP. Initially, the reliable pixels I and the parsing features F are used to instantiate the dense graph nodes V_0, as follows:

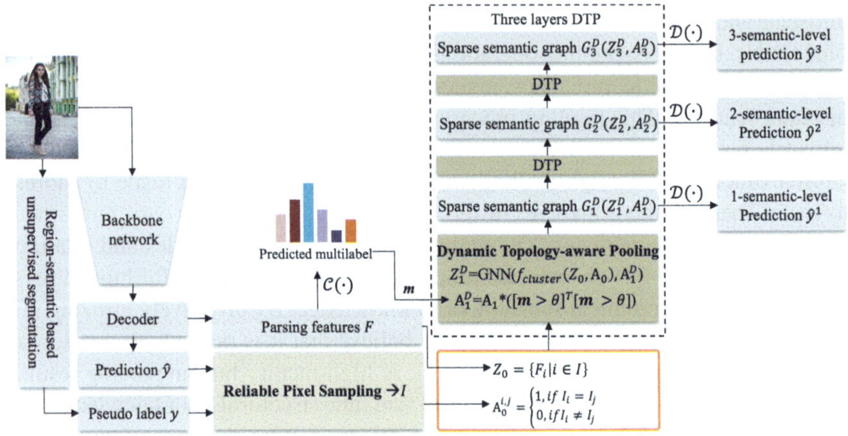

Fig. 6.12 Dynamic topology-aware graph pooling for human parsing

$$V_0 = \{F_i | i \in I\}, \tag{6.20}$$

where I is the set of reliable pixels, F represents the extracted high-dimensional feature of the backbone, and F_i stands for the feature vector of a reliable pixel. All of the reliable pixels will form the dense graph nodes $V_0 \in \mathbb{R}^{d \times C}$ based on the high-dimensional feature F. After this, the dense adjacency matrix A_0 is computed by the pseudo labels:

$$A_0^{i,j} = \begin{cases} 1, & if\ I_i = I_j \\ 0, & if\ I_i \neq I_j, \end{cases} \tag{6.21}$$

where I_i and I_j are the pseudo labels of reliable pixels i and j respectively. The generated V_0 and A_0 instantiate the pixel-level dense graph $G_0(V_0, A_0)$. Since we only apply the reliable pixels to construct the dense graph, it can guarantee the precision of G_0 and suppress the negative impact of noisy labels in the following graph pooling. Concretely, we design a dynamic topology-aware pooling collaborating with multilabel classification which can explicitly encode the highly structured hierarchy of human body into the graph pooling process. Here we detailedly describe the calculation process of the first layer of DTP for simple.

The first layer DTP ($G_0 \rightarrow G_1^D$). DTP utilizes multilabel classification and topology constraints to dynamically pool a dense graph G_0 into a dynamic semantic graph G_1^D, which differs from the static semantic graph G_1. In contrast to G_1, the nodes in G_1^D are flexibly adjusted based on the training samples. The step-by-step procedure is outlined below. Firstly, we employ the multilabel classification function $C(\cdot)$ to generate a probability vector $P_m \in \mathbb{R}^{1 \times N}$, predicting the probabilities for each label. The process is described as follows:

$$P_m = C(F) = \mathcal{S}(f_{2\_\text{MLP}}(\text{GMP}(F))), \tag{6.22}$$

where $f_{2\_MLP}(\cdot)$ represents a two-layer MLP, GMP$(\cdot)$ refers to global max pooling [31], and $\mathcal{S}(\cdot)$ represents the sigmoid function. Then, we dynamically recalibrate the static adjacency matrix A_1 based on the probability vector P_m. This process can produce the dynamic adjacency matrix A_1^D for the dynamic semantic graph G_1^D. As follows:

$$A_1^D = \mathrm{A}_1 * ([P_m > \theta]^T [P_m > \theta]), \tag{6.23}$$

where θ represents a pre-defined threshold, $*$ represents the element-wise multiplication. The notation $[\cdot]$ indicates that the values in P_m will be set to 1 if $P_m > \theta$, and otherwise set to 0. Afterward, we generate the dynamic semantic nodes V_1^D at the 1-semantic level using the dense nodes V_0 and the dense adjacency matrix A_0:

$$V_1^D = \mathrm{GNN}(f_{cluster}(V_0, A_0), A_1^D), \tag{6.24}$$

where $f_{cluster}(\cdot)$ [30] clusters the dense nodes V_0 into the dynamic sparse semantic clusters, and then GNN$(\cdot)$ is used to propagate the information based on the dynamic adjacency matrix A_1^D. By applying this approach, DTP is effectively compelled to pool the dense nodes V_0 and transform them into dynamic semantic nodes $V_1^D \in \mathbb{R}^{N^{(1)} \times C^{(1)}}$. Here, $N^{(1)}$ represents the number of graph nodes at the 1-semantic level, while $C^{(1)}$ denotes the feature dimension. Through this process, DTP gains the ability to selectively activate the graph nodes in the static semantic graph G_1 and generate the dynamic semantic graph G_1^D. The second layer of DTP ($G_1^D \rightarrow G_2^D$) follows the same graph pooling procedure as the first layer, resulting in the generation of $V_2^D \in \mathbb{R}^{N^{(2)} \times C^{(2)}}$. Moving on to the third layer of DTP, the semantic nodes are pooled together into a single node, as described below:

$$V_3^D = f_{cluster}(V_2^D, A_2^D) \in \mathbb{R}^{N^{(3)} \times C^{(3)}}. \tag{6.25}$$

Hierarchical human parsing is achieved through DHSGNN, which employs a three-layer DTP to capture the hierarchical graph representation of the human body, denoted as $V = \left\{V_1^D, V_2^D, V_3^D | V_n^D \in \mathbb{R}^{N^n \times C^n}\right\}$. Following the approach used in Graphonomy [32], the graph representation V_n^D is mapped to feature representation using a trainable projection matrix:

$$F_n^D = \Phi(V_n^D, W_n) \in \mathbb{R}^{h \times w \times c}, n = 1, 2, 3, \tag{6.26}$$

where $\Phi(\cdot)$ represents the projection function used to map the graph representation V_n^D to the same dimension as the feature maps F. The trainable parameter W_n is involved in this mapping process. Ultimately, the output enhanced features are:

$$F_n^{enh} = \mathrm{conv}(\mathrm{cat}(F, F_n^D)), n = 1, 2, 3. \tag{6.27}$$

Acknowledging the significance of spatial information in obtaining accurate human parsing results, we combine the enhanced features obtained from the previous step with the low-level features F_{low}, derived from ResNet-101. This combination is utilized to predict the hierarchical human parsing:

$$\hat{y}^n = \mathcal{D}(F_n^{enh}, F_{low}), n = 1, 2, 3. \tag{6.28}$$

Here, the decoder module is denoted by $\mathcal{D}(\cdot)$, and $\hat{y}^n$ represents the predictions of different semantic levels. Specifically, $\hat{y}^1$ corresponds to the human parsing prediction at the 1-semantic level, $\hat{y}^2$ corresponds to the human parsing prediction at the 2-semantic level, and $\hat{y}^3$ corresponds to the human parsing prediction at the 3-semantic level. The decoder module typically consists of an upsampling layer and three convolution layers. The upsampling layer employs the bilinear interpolation algorithm for upsampling the feature maps.

6.3.4 Experimental Results and Discussions

We introduce a novel graph pooling called Dynamic Topology-aware Pooling (DTP), which efficiently utilizes semantic structural prior knowledge to generate a dynamic hierarchical semantic graph. This graph construction aids in learning weakly supervised human parsing. In this chapter, we apply three layers of DTP to build a dynamic hierarchical semantic graph. To assess the effectiveness of DTP, we conduct several ablation experiments on various dynamic hierarchical semantic graphs. The results of these experiments are presented in Table 6.8. Specifically, they denote the first layer of DTP as DTP_1, the second layer as DTP_2, and the third layer as DTP_3. Notably, the model achieves the best results, with a mIoU of 57.05%, by adopting the three layers of DTP on the ATR test set. This finding indicates that modeling the semantic structural prior knowledge of the human body in an optimal way can significantly contribute to representation learning in weakly supervised human parsing. In conclusion, the Dynamic topology-aware pooling approach, with its dynamic hierarchical semantic graph construction, proves to be highly effective in learning from weakly supervised data for accurate human parsing. By leveraging the semantic

Table 6.8 The effect of dynamic topology-aware pooling is evaluated on ATR test [16]

DTP_1	DTP_2	DTP_3	mIoU	Pixel accuracy	F1-score	Recall
–	–	–	36.41	84.14	52.52	50.38
✓	–	–	45.12	89.04	59.85	59.20
✓	–	✓	53.15	91.21	67.79	63.51
✓	✓	-	54.88	91.89	67.72	64.39
✓	✓	✓	**57.05**	**92.75**	**69.89**	**66.12**

structural prior knowledge efficiently, the model achieves a well results and provides valuable insights into representation learning for weakly supervised human parsing tasks.

The results of the ablation experiments on the dynamic hierarchical semantic graphs confirm the significance of each layer of DTP in improving the model's performance. Delating any layer of the dynamic semantic graph leads to a degradation in model performance. Particularly, when only adopting the first layer of DTP (baseline+DTP_1), the model achieves a relatively worse mIoU of 45.12% compared to baseline+DTP_1+DTP_2 and baseline+DTP_1+DTP_3. To visually demonstrate the effectiveness of the dynamic hierarchical semantic graph generated by DTP, this chapter presents some visual results in Fig. 6.13. The baseline model exhibits poor predictions with serious errors, such as failing to segment the legs, pants, and small objects like hats and belts. However, with the aid of the hierarchical semantic graph generated by the first layer of DTP (baseline+DTP_1), these serious errors are significantly improved, as shown in Fig. 6.13a. Moreover, as additional layers of DTP are added (Fig. 6.13b and c), the predictions continue to improve gradually. This visual results supports the conclusion that a single semantic graph is no longer sufficient

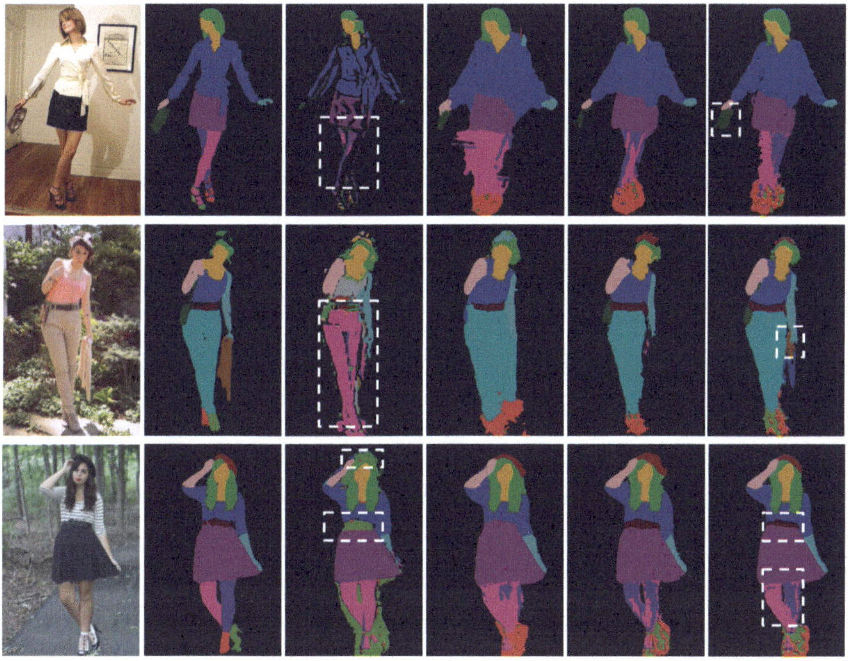

Fig. 6.13 Visualization results of the ablation experiments for dynamic topology-aware pooling on the ATR test [16]. **a** Visualization results of baseline+DTP_1. **b** Visualization results of baseline+DTP_1+DTP_2. **c** Visualization results of baseline+DTP_1+DTP_2+DTP_3

for addressing the challenges of weakly supervised human parsing. As previously analyzed, the dynamic hierarchical semantic graph generated by DTP allows model to capture rich hierarchical structural features, leading to improved performance in weakly supervised human parsing. Notably, the best results are achieved by model when using three layers of DTP (baseline+DTP_1+DTP_2+DTP_3). In Fig. 6.13c, the white dashed rectangle demonstrates the superior performance of proposed model in precisely handling small parts like bags and belts. This highlights the capability of the proposed DTP to accurately capture fine-grained details. Additionally, the dynamic hierarchical semantic graph generated by three layers of DTP enables the model to learn more accurate structural features, enabling model to distinguish between the left and right legs, which was a challenging task for baseline+DTP_1 and baseline+DTP_1+DTP_2. In summary, the quantitative experimental results in Table 6.8 and visualizations in Fig. 6.13 clearly demonstrate the critical role of the dynamic hierarchical semantic graph generated by DTP in improving the model's performance. The proposed DTP proves to be an effective solution for capturing the complex structural relationships and semantic features of human body parts, leading to significant improvements in weakly supervised human parsing results.

In human parsing task, human bodies present a natural topological physical structure, and how to effectively use the specific structure to construct the optimal graph is an essential avenue to explore. In this chapter, we proposed a new Multilabel Learning based Adaptive Graph Convolutional Network (ML-AGCN) for human parsing, which can generate adaptive graphs according to the content of the image. Compared with previous graph-based methods, the ML-AGCN can effectively handle the problem that the optimal graph for each training sample varies in both size and connectivity. From a new perspective, we use the adaptive matrix generated by multilabel learning to achieve GCA and GNA and then generate an adaptive graph, which can remedy the weakness of the fixed graph and tackle the problem of occlusions in human parsing. Besides, we attempt to sufficiently fuse fixed graph and adaptive graph for a fusion graph with rich information. Our method, which comprehensively utilizes fixed graph, adaptive graph and fusion graph to obtain accurate and rich features, achieves a superior performance on single-person human parsing as well as multi-person human parsing. However, it only applies a simple multilabel classification model to verify the feasibility of multilabel classification in generating adaptive graph for human parsing. In spite of achieving good performance on four challenging datasets, it performs poorly when parsing the heterogeneous data with many labels. As a result, we will try to use heterogeneous graph network to perform human parsing on the datasets with more heterogeneous labels in the future.

In this chapter, we aim to research the dynamic graph neural networks for human parsing. Motivated by this, we propose dynamic graph construction for human parsing and dynamic graph pooling for human parsing. Specially, we propose a new Multilabel Learning-based Adaptive Graph Convolutional Network (ML-AGCN) for human parsing, which generates adaptive graphs based on the content of the image. Unlike previous graph-based methods, ML-AGCN effectively handles the varying size and connectivity of optimal graphs for each training sample. By using an adaptive matrix generated through multilabel learning, we achieve Graph Convolutional

Attention (GCA) and Graph Node Attention (GNA) to create an adaptive graph. This approach addresses the limitations of fixed graphs and tackles occlusion issues in human parsing. Our method achieves superior performance in both single-person and multi-person human parsing. Further, we propose a dynamic graph pooling method (Dynamic Topology-aware Pooling, DTP) for human parsing. This enables the model to dynamically adjust the receptive fields during the pooling process, which is crucial for capturing both local and global context in an image. The experiments demonstrates that DTP can fully leverage the inherent structure of the human body, resulting in enhanced parsing performance.

References

1. Li, R., Wang, S., Zhu, F., Huang, J.: Adaptive graph convolutional neural networks. In: Proceedings of the AAAI Conference on Artificial Intelligence, pp. 3546–3553 (2018)
2. Wang, X., Zhu, M., Bo, D., Cui, P., Shi, C., Pei, J.: Am-gcn: adaptive multi-channel graph convolutional networks. In: Conference on Knowledge Discovery and Data Mining, pp. 1243–1253 (2020)
3. Shi, L., Zhang, Y., Cheng, J., Lu, H.: Two-stream adaptive graph convolutional networks for skeleton-based action recognition. In: Proceedings of the IEEE Conference on Computer Vision and Pattern Recognition, pp. 12026–12035 (2019)
4. Gong, K., Gao, Y., Liang, X., Shen, X., Wang, M., Lin, L.: Graphonomy: universal human parsing via graph transfer learning. In: Proceedings of the IEEE Conference on Computer Vision and Pattern Recognition, pp. 7450–7459 (2019)
5. He, H., Zhang, J., Zhang, Q., Tao, D.: Grapy-ml: graph pyramid mutual learning for cross-dataset human parsing. In: Proceedings of the AAAI Conference on Artificial Intelligence, pp. 10949–10956 (2020)
6. Chen, X., Li, Z., Yuan, Y., Yu, G., Shen, J., Qi, D.: State-aware tracker for real-time video object segmentation. In: Proceedings of the IEEE Conference on Computer Vision and Pattern Recognition, pp. 9381–9390 (2020)
7. Cao, J., Anwer, R.M., Cholakkal, H., Khan, F.S., Pang, Y., Shao, L.: Sipmask: spatial information preservation for fast image and video instance segmentation. In: Proceedings of the European Conference on Computer Vision, vol. 12359, pp. 1–18 (2020)
8. Liang, X., Gong, K., Shen, X., Lin, L.: Look into person: joint body parsing & pose estimation network and a new benchmark. IEEE Transactions on pattern analysis and machine intelligence 41(4), 871–885 (2019)
9. Zhang, Z., Su, C., Zheng, L., Xie, X.: Correlating edge, pose with parsing. In: Proceedings of the IEEE Conference on Computer Vision and Pattern Recognition, pp. 8897–8906 (2020)
10. Wang, W., Zhang, Z., Qi, S., Shen, J., Pang, Y., Shao, L.: Learning compositional neural information fusion for human parsing. In: IEEE International Conference on Computer Vision, pp. 5702–5712 (2019)
11. Gong, K., Liang, X., Li, Y., Chen, Y., Yang, M., Lin, L.: Instance-level human parsing via part grouping network. In: Proceedings of the European Conference on Computer Vision, vol. 11208, pp. 805–822 (2018)
12. Nie, X., Feng, J., Yan, S.: Mutual learning to adapt for joint human parsing and pose estimation. In: Proceedings of the European Conference on Computer Vision, vol. 11209, pp. 519–534 (2018)
13. He, K., Zhang, X., Ren, S., Sun, J.: Deep residual learning for image recognition. In: Proceedings of the IEEE Conference on Computer Vision and Pattern Recognition, pp. 770–778 (2016)

14. Zhao, H., Shi, J., Qi, X., Wang, X., Jia, J.: Pyramid scene parsing network. In: Proceedings of the IEEE Conference on Computer Vision and Pattern Recognition, pp. 6230–6239 (2017)
15. Chen, X., Mottaghi, R., Liu, X., Fidler, S., Urtasun, R., Yuille, A.L.: Detect what you can: detecting and representing objects using holistic models and body parts. In: Proceedings of the IEEE Conference on Computer Vision and Pattern Recognition, pp. 1979–1986 (2014)
16. Liang, X., Xu, C., Shen, X., Yang, J., Liu, S., Tang, J., Lin, L., Yan, S.: Human parsing with contextualized convolutional neural network. In: Proceedings of the IEEE Conference on Computer Vision and Pattern Recognition, pp. 1386–1394 (2015)
17. Liang, X., Gong, K., Shen, X., Lin, L.: Look into person: joint body parsing & pose estimation network and a new benchmark. IEEE Transactions on pattern analysis and machine intelligence 41(4), 871–885 (2019)
18. Gong, K., Liang, X., Li, Y., Chen, Y., Yang, M., Lin, L.: Instance-level human parsing via part grouping network. In: Proceedings of the IEEE Conference on Computer Vision and Pattern Recognition, pp. 805–822 (2018)
19. Zhang, X., Chen, Y., Zhu, B., Wang, J., Tang, M.: Part-aware context network for human parsing. In: Proceedings of the IEEE Conference on Computer Vision and Pattern Recognition, pp. 8968–8977 (2020)
20. Wang, W., Zhang, Z., Qi, S., Shen, J., Pang, Y., Shao, L.: Learning compositional neural information fusion for human parsing. In: IEEE International Conference on Computer Vision, pp. 5702–5712 (2019)
21. Wang, W., Zhou, T., Qi, S., Shen, J., Zhu, S.: Hierarchical human semantic parsing with comprehensive part-pelation modeling. IEEE Transactions on pattern analysis and machine intelligence pp. 1–15 (2021)
22. Luo, Y., Zheng, Z., Zheng, L., Guan, T., Yu, J., Yang, Y.: Macro-micro adversarial network for human parsing. In: Proceedings of the European Conference on Computer Vision, pp. 424–440 (2018)
23. Fang, H., Lu, G., Fang, X., Xie, J., Tai, Y., Lu, C.: Weakly and semi supervised human body part parsing via pose-guided knowledge transfer. In: Proceedings of the IEEE Conference on Computer Vision and Pattern Recognition, pp. 70–78 (2018)
24. Chen, L., Zhu, Y., Papandreou, G., Schroff, F., Adam, H.: Encoder-decoder with atrous separable convolution for semantic image segmentation. In: Proceedings of the IEEE Conference on Computer Vision and Pattern Recognition, pp. 833–851 (2018)
25. Li, P., Xu, Y., Wei, Y., Yang, Y.: Self-correction for human parsing. IEEE Transactions on pattern analysis and machine intelligence pp. 1–12 (2020)
26. Gong, K., Liang, X., Zhang, D., Shen, X., Lin, L.: Look into person: self-supervised structure-sensitive learning and a new benchmark for human parsing. In: Proceedings of the IEEE Conference on Computer Vision and Pattern Recognition, pp. 6757–6765 (2017)
27. Zhang, Z., Su, C., Zheng, L., Xie, X.: Correlating edge, pose with parsing. In: Proceedings of the IEEE Conference on Computer Vision and Pattern Recognition, pp. 8897–8906 (2020)
28. Chen, L., Papandreou, G., Kokkinos, I., Murphy, K., Yuille, A.L.: Deeplab: semantic image segmentation with deep convolutional nets, atrous convolution, and fully connected crfs. IEEE Transactions on pattern analysis and machine intelligence 40(4), 834–848 (2018)
29. Ruan, T., Liu, T., Huang, Z., Wei, Y., Wei, S., Zhao, Y.: Devil in the details: towards accurate single and multiple human parsing. In: Proceedings of the AAAI Conference on Artificial Intelligence, pp. 4814–4821 (2019)
30. Yuan, Y., Chen, X., Wang, J.: Object-contextual representations for semantic segmentation. In: Proceedings of the European Conference on Computer Vision, pp. 173–190 (2020)
31. Li, T., Liang, Z., Zhao, S., Gong, J., Shen, J.: Self-learning with rectification strategy for human parsing. In: Proceedings of the IEEE Conference on Computer Vision and Pattern Recognition, pp. 9260–9269 (2020)
32. Luo, X., Su, Z., Guo, J., Zhang, G., He, X.: Trusted guidance pyramid network for human parsing. In: Proceedings of the ACM International Conference on Multimedia, pp. 654–662 (2018)

Chapter 7
Graph Neural Networks for Video Object Segmentation

As one of the key methods of scene understanding, video object segmentation can be used in applications such as video conferencing, video editing and autonomous driving. Recently, many video object segmentation methods have been proposed. In this chapter, we extend graph neural networks to video object segmentation to enhance the utilization of target structure information. We describe in detail the video object segmentation methods based on graph neural networks in terms of graph construction, positional encoding for graph structure, and graph convolution operations.

7.1 Introduction

Video object segmentation has a wide range of applications, such as object tracking, video conferencing, action recognition, autonomous driving, etc. In recent years, more semi-supervised video object segmentation algorithms have been proposed. However, existing algorithms do not fully utilize the structural features of the object. These methods usually use convolutional networks or Transformer modules to process the whole image, and extract the features of the image through multi-layer network processing, without fully considering the structural features of the target, such as the shape and structural information of the rod, rope, and other objects. We extend graph neural networks to video object segmentation to enhance the utilization of target structure information.

A large amount of graph data exists in everyday life, such as social networks and recommendation systems. Graph data processing and graph representation learning are important areas of machine learning. In recent years, graph neural networks have been widely used and have achieved great success in areas such as

W. Liu et al., *Graph Neural Network Methods and Applications in Scene Understanding*,
https://doi.org/10.1007/978-981-97-9933-6_7

recommendation systems [1], hand pose estimation [2]. The development of graph neural networks have greatly advanced the research progress of graph representation learning, demonstrating powerful feature extraction capabilities on graph data. Several researchers [3–5] have introduced graph neural networks into video object segmentation. AGNN [3, 4] constructs a fully connected graph by treating video frame features as nodes. AGNN explores rich and higher-order inter-frame relationships through iterative information fusion. EGMN [5] enhances AGNN by introducing learnable controllers. The controllers are used for memory reading and writing. The encoded features of the history frames with their prediction masks are treated as nodes of the fully connected graph. However, the way these methods construct graph structures lacks interpretability and does not take full advantage of the feature extraction capabilities of graph neural networks.

Recently, Song et al. [6] propose a Learnable Tree Filter (LTF). The authors use the Boruvka's algorithm to generate a minimum spanning tree, calculate the similarity weight between adjacent nodes based on node features, and then accumulate the weights of all connected edges on the path of two non-adjacent nodes to obtain the similarity weight between that pair of non-adjacent nodes. LTF can effectively capture the structural features of objects. Hou et al. [7] propose the strip pooling method. The authors conclude that global average pooling ignores the shape structural information of the target and therefore use horizontal and vertical strip pooling with better results. Different from the above methods, this chapter utilizes an agglomerative clustering algorithm to generate a series of sub-minimum spanning trees, and then uses graph convolutional networks for structural feature extraction. Graph convolutional networks have recently been of interest to many researchers and have powerful learning and feature extraction capabilities for non-Euclidean data. In this chapter, we use a coalescent clustering algorithm to obtain a series of sub-minimal spanning trees, which are the generated graph structures. Each sub-minimum spanning tree corresponds to an image region. The generated graph structure is disconnected at the object boundary and connected within the region, effectively capturing the object structure information. The clustering operation in the feature space also allows for better robustness in the graph structure construction process. The object structure features are then further enhanced using a graph convolutional network.

7.2 Graph Construction of Video Data

This section describes graph construction for video data. First, we review existing graph construction methods for video object segmentation. Then, the proposed minimum spanning tree based graph structure construction process is described. Furthermore, we adjust the generated graph structure according to the object region of the video frame, breaking the graph structure at the object boundary to enhance the structural features of the object region.

7.2.1 Graph Construction Based on Minimum Spanning Tree

The most intuitive method is to define the four-neighborhood or eight-neighborhood of a pixel point to get the graph structure of the video frame, and further the segmentation result of the current video frame can be obtained by algorithmic processing of the node or edge features to predict the labels of the output edges, as shown in Fig. 7.1. The graph structure of the whole video can be constructed by establishing temporal connections based on the 3D coordinates of pixels in the video or optical flow. The traditional methods are based on constructing the graph structure, designing a certain energy function and predicting the node labels or edge labels to realize the video object segmentation, and these methods are known as graph structure based video object segmentation methods. Such methods cannot effectively deal with challenges such as fast motion, scale changes, etc. Researchers have started to introduce deep learning to graph structure based video object segmentation methods. Literature [8] employs convolutional neural networks to model the interdependence of nodes. Gradually, some approaches introduce graph neural networks to video object segmentation.

Wang et al. [3] achieve video object segmentation by constructing fully connected graphs. Each video frame in a video sequence is regarded as a graph node, as shown in Fig. 7.2. The rich and higher-order relationships between video frames are mined through the message passing process of graph neural networks. The constructed graph is a directed graph, with two edges in different directions between two nodes, and each node has self-connected edges. To preserve spatial detail, the node features are tensors in three dimensions, obtained using the DeepLabV3 backbone network. Self-connected edge embedding features are obtained through the self-attention mechanism [9, 10] and are three-dimensional tensors. Inter-frame edges are obtained

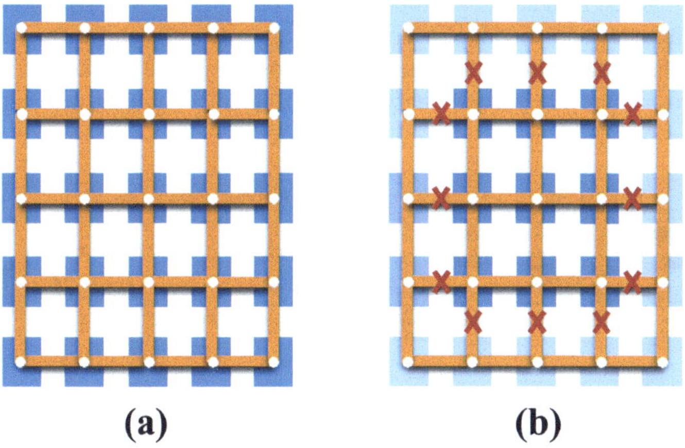

Fig. 7.1 **a** Graph structure of a video frame defined by four-neighborhoods, **b** Segmentation result of a video frame represented by the graph structure

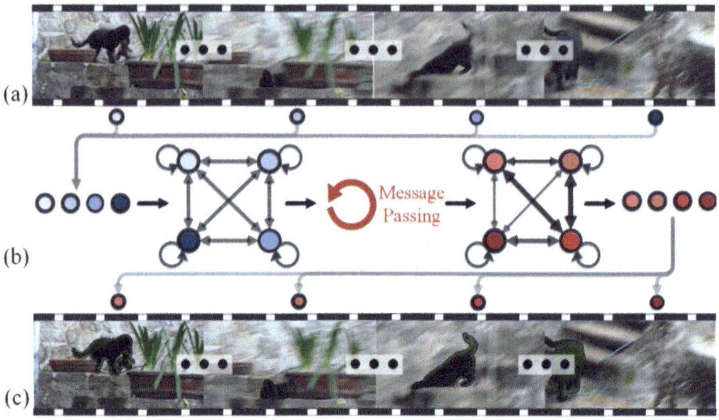

Fig. 7.2 Graph constructed by EGMN [5]

through co-attention [11], forming two-dimensional matrices. The authors add a gating mechanism to the message aggregation process to enhance the robustness of the aggregation process. Node state updates are then completed using ConvGRU.

Wang et al. propose EGMN [5] based on AGNN [3, 4]. EGMN designs graph memory networks, constructs fully connected graphs, and uses video frames as nodes, as shown in Fig. 7.3. Unlike AGNN, the number of nodes in the graph constructed by EGMN is fixed. The first frame containing the annotations and the segmented history frames form the support set, and EGNN samples a fixed number of video frames from the support set as graph nodes. The video frames with their masks are fed into the backbone network to extract node features. In addition, EGMN introduces learnable read and write controllers during each message pass to efficiently update the target appearance representation through meta-learning methods.

In recent years, Hou et al. [7] propose a strip pooling method. Traditional spatial pooling methods such as global average pooling ignore the shape structure information of the target. The authors adopt strip pooling in the vertical or horizontal direction to capture the banded shape and can model long-term dependencies. A

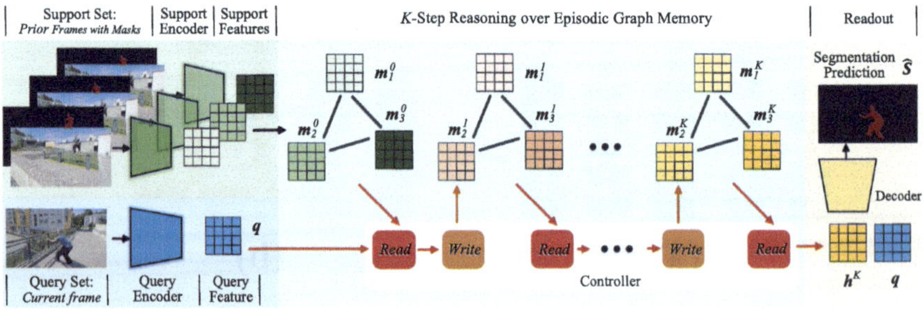

Fig. 7.3 Graph constructed by AGNN [3]

Learnable Tree Filter (LTF) is designed by Song et al. [6]. LTF employs the Boruvka algorithm to generate a minimum spanning tree. Then, the original feature is adjusted by the cumulative weight on the path of non-adjacent nodes and the learnable unary weight. Unlike LTF, after using the minimum spanning tree to obtain the graph structure, we utilize a graph convolutional neural network to process the graph data composed the graph structure and video frame features. The object structural features are effectively enhanced.

The Minimum Spanning Tree (MST) is a subset of the initial graph structure edge set E and connects all nodes with the smallest sum of weights. Kruskal and Prim algorithms are commonly used to solve the Minimum Spanning Tree. Kruskal sorts all edges by weight from smallest to largest and then selects in turn the edges that do not form loops with the edges already added to MST. Prim divides all nodes into the set of spanning tree nodes and the set of unvisited nodes, takes any node as the initial node of the Spanning tree, and then successively adds the node with the least weight of the connecting edge between the unvisited node and the Spanning tree node to the set of spanning tree nodes. Brokva algorithm combines the advantages of these two algorithms. At each iteration, Brokva first traverses all the edges to find the least weighted connected edge between each subtree and the other subtrees, then traverses the resulting set of least weighted connected edges and merges the different subtrees connected by the least connected edge.

Specifically, the embedding features of the video frames are extracted using the backbone network. The feature at each position of the video frame embedding features is a vector, i.e., a graph node feature. If four neighbours are used, the graph node at each position in the initial graph structure connects its four neighbouring graph nodes above, below, left and right. If an eight neighbours are used, the graph node at each position in the initial graph structure connects its upper, lower, left, right and eight neighbouring graph nodes, upper left, lower left, upper right and lower right. The Euclidean distance between node features is used as the weight of edges. The steps of Brokva algorithm to find the Minimum Spanning Tree on the initial graph structure are as follows:

Step 1: Initialize each graph node as an independent subtree.

Step 2: If the number of subtrees is greater than 1, repeat steps 3 and 4.

Step 3: Traverse all edges of the initial graph structure to find the root nodes of the subtrees to which the two nodes of the connected edge belong. If the root nodes of the two nodes of the connected edge do not belong to the same subtree, and the weight of the current connected edge is less than the edge weight of the connected edge stored in the root node marker variable, the root node marker variable is updated to the edge ordinal number of the current connected edge.

Step 4: Traverse all nodes of the initial graph structure, and for nodes that update the node marker variable, if the root nodes of the two nodes of the edge stored in the node marker variable do not belong to the same subtree, the two subtrees connected by the current edge will be merged.

Step 5: Output the Minimum Spanning Tree.

After a round of subtree merging, the number of subtrees is at least halved, so the number of loops in step 2 is at most $log(V)$. The complexity of step 3 is at most $O(E)$. The complexity of subtree merging in step 4 is constant, so the total complexity is approximately $O(Elog(V))$. The algorithm complexity can be further reduced by using methods such as dictionaries and weight divide and conquer.

7.2.2 Graph Refinement According to Object Regions

Figure 7.4 shows the graph structure generated by the algorithm proposed in this section, where eight directly adjacent neighbours nodes of each node are considered in the initial graph structure. When the initial graph structure uses four neighbours, some edges of the generated graph structure cross the region boundaries, i.e. nodes with dissimilar features are connected. Moreover, in order to simultaneously enhance the structural features within the target region, an agglomerative clustering method is used to generate the graph structure, so that the graph structure breaks at the object boundaries.

For each location the feature points connect their top, bottom, left and right and top-left, bottom-left, top-right and bottom-right eight neighbourhood feature points to construct the initial graph structure $G = (V, E)$. V denotes the set of feature points

Fig. 7.4 The graph structure output by the graph structure construction module

at all locations of the video frame, i.e. the node set of the graph structure. E denotes the connected edges of the initial graph structure. Edge weights are then calculated using negative squared distances and a series of sub-minimum spanning trees are generated using a aggregation clustering algorithm. The sub-minimum spanning trees form a graph structure that is disconnected at object boundaries and connected within object regions, thus effectively capturing the target structural information. A segmentation result P of a video frame corresponds to a regional partition of the video frame, which can be expressed as $G' = (V, E')$, where $E' \in E$. That is, the set of edges corresponding to any partition result is a subset of E.

Specifically, a sub-minimum spanning tree corresponds to an image region, $R \in V$. The intra-region difference $\mathrm{Int}(R)$ is the maximum of all connected edge weights of the sub-minimum spanning tree $\mathrm{MST}(R, E)$ within that region, as shown in the following equation.

$$\mathrm{Int}(R) = \max_{e \in \mathrm{MST}(R,E)} w(e). \tag{7.1}$$

Further, the inter-region difference is defined as the minimum edge weight connecting the two regions R_1 and R_2, as shown in Eq. (7.2). If there are no connecting edges between two regions R_1 and R_2, we let $\mathrm{Dif}(R_1, R_2) = \infty$.

$$\mathrm{Dif}(R_1, R_2) = \min_{v_i \in R_1, v_j \in R_2, (v_i, v_j) \in E} w\left(\left(v_i, v_j\right)\right). \tag{7.2}$$

If the minimum value in Eq. (7.2) is replaced with a median or other value, the problem of finding a good region partition is NP-hard [12]. The proposed method adopts the following rule to decide whether to merge two regions: if the difference between two regions is greater than the intra-regional difference in at least one of the two regions then the two regions will not be merged, as shown in the following equation.

$$D(R_1, R_2) = \begin{cases} \text{true} & \text{if } \mathrm{Dif}(R_1, R_2) > \mathrm{MInt}(R_1, R_2) \\ \text{false} & \text{otherwise}, \end{cases} \tag{7.3}$$

where the minimum internal difference MInt is defined as follows:

$$\begin{gathered} \mathrm{MInt}(R_1, R_2) = \\ \min\left(\mathrm{Int}(R_1) + \gamma(R_1), \mathrm{Int}(R_2) + \gamma(R_2)\right). \end{gathered} \tag{7.4}$$

The threshold function γ controls the extent to which the inter-regional difference $D(R_1, R_2)$ is greater than the intra-regional difference $\mathrm{Int}(R_1)$ or $\mathrm{Int}(R_2)$. It is because for small regions, $\mathrm{Int}(R)$ cannot accurately estimate the local attributes of the data. In the extreme case, when $|R| = 1$, $\mathrm{Int}(R) = 0$, each point becomes a separate region. Therefore, a threshold function $\gamma(R)$ based on the size of the region is employed, as shown in the following equation:

$$\gamma(R) = c/|R|, \tag{7.5}$$

where $|R|$ denotes the number of pixels in the region R and c is a constant. That is, for small regions, a threshold function is needed to increase its intra-regional difference. A larger c tends to give a larger size to the region obtained by clustering. However, c is not the minimum region size, as many small regions still be formed when the local inter-regional difference is large. γ can use any non negative function regarding a single region without affecting the convergence of the clustering algorithm. Different γ can be set to obtain regions of different shapes, e.g. using the ratio of perimeter to area causes the resulting elongated regions to be merged.

The following specifies the tree generation process, which is similar to Kruskal's algorithm used to generate minimum spanning trees. Unlike Kruskal's algorithm, the proposed method adopts different edge decision rules and outputs multiple minimum spanning trees. The algorithm is implemented with a computational complexity of $O(n \log n)$, where n is the number of edges in the initial graph G. The input to the algorithm is an initial graph G composed within the feature space of a video frame, with the number of nodes and edges m and n respectively. The output of the algorithm is a series of sub-minimum spanning trees that divide the node V into different regions $P = (R_1, R_2, \ldots, R_s)$.

Step 1: Arrange the boundary sets E in ascending order by weight, $\pi = (b_1, b_2, \ldots, b_n)$.

Step 2: The initial division is P^0, with each node being a region.

Step 3: Repeat step 4 for $q = 1, 2, \ldots, n$.

Step 4: Given P^{q-1}, the proposed construct P^q using the following procedure. Let the qth edge b_q after sorting connects nodes v_i and v_j. R_i^{q-1} denotes the region of P^{q-1} that contains node v_i and R_j^{q-1} denotes the region that contains node v_i. If $R_i^{q-1} \neq R_j^{q-1}$ and $w\left(b_q\right) \leq Mint\left(R_i^{q-1}, R_j^{q-1}\right)$, then merging R_i^{q-1} and R_j^{q-1} in P^{q-1} yields P^q. Otherwise, $P^q = P^{q-1}$.

Step 5: Output $P = P^n$.

The above steps lead to a series of sub-minimum spanning trees corresponding to a region partition result. It can be shown that the above algorithm yields the same region partition result even if there are edges with equal edge weights [12].

Figure 7.4 shows the output of the graph structure construction module of the proposed method, with a selection of video frames from each of the three video sequences. The graph structures in some small regions of the figure are omitted for presentation purposes. From these results, it can be seen that the output graph structures are connected inside the target and background regions, corresponding to different sub-minimum spanning trees that can effectively capture the structural information inside the target region, and are disconnected at the boundaries of the different regions, separating the different sub-minimum spanning trees. For example, the boundary between the camel and the background region in the first video, the boundary between the quad bike and the background region in the third video, and the boundary between the person carrying something and the background region in the fourth video. Using the obtained graph structure it is possible to implement enhancement of features that lie on the same sub-minimum spanning tree, thus enhancing the structural features of the target.

7.3 Integrating Spatial Graph Convolution and Transformer

Based on the graph construction, we employ a graph convolutional network to process the graph data consisting of the graph structure and video frame features to further enhance the structural features of the target. The structural information of the nodes is first encoded using positional encoding techniques, and the encoded structural information is concatenated with the node features to obtain the enhanced node features. Further, we draw on recent advances in Transformer methods for video object segmentation. For the first time, we combine graph convolutional networks with Transformer methods in the field of video object segmentation, achieving higher segmentation performance.

7.3.1 Positional Encoding for Graph Structure

With the increasing popularity of smart devices such as mobile phones and tablets, images and videos have become important media for recording and communicating information. An image is a bitmap made up of an array of pixel dots, which is a typical Euclidean data. Pixel position information is an important cue for visual understanding. In dense prediction tasks such as object detection and instance segmentation, position information is particularly important. Social network data, chemical composition data, biological protein data and knowledge graph data are non-Euclidean structured data and graphs are an important way of describing these data.

In non-Euclidean structured data, data points are usually irregularly arranged and the number of neighbor nodes connected to different data points is usually different, which is very different from Euclidean structured data. In addition to node location information, local structure information of nodes is an important clue for graph data processing tasks, especially node-level tasks. So for graph data, the position encoding should not only consider the location information of the graph nodes but also the structural features of the graph nodes. We design a graph structure generation algorithm based on agglomerative clustering to capture the structural features of objects in video frames. Further, we use a graph convolutional network to enhance the object structure features.

Laplace Positional Encoding (LapPE) and Random Walk Positional Encoding (RWPE) are two commonly used positional encoding methods in graph representation learning. Given a graph data $G = (A, X)$, the positional encoding matrix $P = \mathrm{PE}(A)$, where each row of P represents the positional encoding vector of each node, also called the node positional feature.

For the Laplace positional encoding [13], the normalized Laplace matrix L is first obtained using Eq. (7.6), where I is the unit matrix and D is the degree matrix. Then the eigenvalues and eigenvectors of the Laplace matrix L are obtained using the eigenvalue decomposition as shown in Eq. (7.7), where Λ is a diagonal matrix with diagonal elements $0 \leq \lambda_1 \leq \lambda_2 \cdots \leq \lambda_N$, the eigenvalues of matrix L. $U = [u_1, u_2, \ldots, u_N]^T$ is an orthogonal matrix with $Lu_i = \lambda_i u_i$ for the i-th eigenvector u_i.

$$L = I - \hat{A} = I - D^{-\frac{1}{2}} A D^{-\frac{1}{2}}, \tag{7.6}$$

$$L = U^T \Lambda U. \tag{7.7}$$

The eigenvectors are normalized to unit length. During training, the signs of the eigenvectors are randomly flipped to learn invariance. Because the smallest eigenvalues have better smoothing, k smallest eigenvalues are used for the final positional encoding, where $k < N$, N denotes the number of the graph nodes. The computational complexity of the above eigenvalue decomposition is $O(E^{(3/2)})$, where E denotes the number of the graph edges. Although the eigenvalue decomposition of Laplace matrix can be accelerated using the Nystrom method [14], it still has a high computational complexity.

For the random walk positional encoding, RWPE is obtained by the diagonal elements of the m-step random walk matrix $W_r = (AD^{-1})^m$, where A and D are the adjacency matrix and degree matrix, respectively. The positional feature p_i^{RWPE} of node i is shown in the following equation:

$$p_i^{RWPE} = \left[\mathrm{RP}_{ii}, \mathrm{RP}_{ii}^2, \ldots, \mathrm{RP}_{ii}^m\right] \in \mathbb{R}^m, \tag{7.8}$$

where $\mathrm{RP} = AD^{-1}$ denotes the random walk operation. Li et al. [15] use all elements of the random walk matrix. LSPE [16] uses only the diagonal elements of the random walk matrix, i.e., RP_{ii}, and thus has a low computational complexity. RP_{ii} represents the probability of node i returning to itself. RP_{ii_m} represents the probability of node i returning to itself after m-step transition. RWPE do not encounter the sign ambiguity issue that LapPE faces, thereby alleviating the need for the network to learn additional invariance. If a node a has a unique m-hop subgraph, RWSE can produce a unique node representation for a sufficiently large m. In the video object segmentation experiment, we set m to 16 based on grid search.

7.3.2 Spatial Graph Convolution Integrated with Transformer

Compared to online learning based methods, matching based methods usually have a faster running speed. Matching-based methods rely on establishing correspondences between the features of the current frame and a reference frame, facilitating transduction inference from reference frame segmentation to the current frame. However, these methods are less suitable for scenarios where the target frequently vanishes due to their reliance on a limited number of reference frames, typically not exceeding two.

In the PLM approach [17], a Siamese network is employed to extract features from both the query and search frames. These features undergo encoding through multiple fully connected layers to establish pixel-level similarities. Meanwhile, PML [17] adopts a K-Nearest-Neighbor (KNN) method for pixel label prediction within the feature space. It's worth noting that KNN can sometimes lead to mis-segmentation, particularly in cases where background objects or regions exhibit similarity. On the other hand, VideoMatch [18] also leverages a Siamese architecture to extract features from template and current frames. It introduces a soft matching mechanism to align these frame features. However, VideoMatch is susceptible to a degree of mismatch, primarily due to the absence of temporal information in its approach. FEELVOS [19] focuses solely on extracting segmentation results from the extreme values within similarity maps, potentially underutilizing valuable information present in these maps. Conversely, RANet [20] introduces a ranking attention module to prioritize and select similarity maps between the features of the initial and current frames. However, RANet's approach of directly concatenating previous frame segmentation results within the feature space may not fully harness the potential of the previous frame's feature information.

Certain techniques in this domain employ feature memories to retain additional video frame features, thereby incorporating valuable temporal context for better target understanding. For instance, STM [21] systematically accumulates features extracted from video frames and their corresponding predictions into a dedicated memory unit, subsequently accessing these memorized features through non-local operations. Furthermore, LCM [22] extends the capabilities of STM [21] by introducing an object relation module, enhancing the model's contextual understanding of objects within the video. RMNet [23] enhances the capabilities of STM [21] by elevating the matching mechanism. It accomplishes this by substituting the original global-to-global matching with a local-to-local matching scheme. On the other hand, GIEL [24] employs a methodology that predicts global foreground information and disentangles distinct instances through global embedding learning and instance embedding learning. To further refine its performance, GIEL introduces a temporal attention mechanism, allowing it to emphasize specific frame features within the support set.

The inception of the Transformer technique can be attributed to Vaswani et al.'s groundbreaking work [9], which achieved remarkable success in the realm of machine translation. Since then, a continuous stream of Transformer-based methodologies has emerged, revolutionizing the landscape of natural language processing techniques and their diverse applications. Notably, the introduction of the Vision Transformer (VIT) approach by Dosovitskiy et al. [25] has led to the widespread adoption of visual Transformers across various domains, including image classification [26], object detection [27], and image segmentation [28], among others. Furthermore, recent years have witnessed the emergence of Transformer-based methods for video object segmentation, marking yet another exciting frontier in this transformative technology. In the SST framework [29], it leverages the Transformer module to capture spatio-temporal context features and create an object similarity tensor. Notably, it incorporates a sparse attention operator, meticulously crafted to minimize computational demands.

On the other hand, JOINT [30] introduces an integrated approach, comprising both transductive and inductive inference branches. For the transductive inference branch, it adopts a modern Transformer architecture, while employing a few-shot learner for the inductive inference branch. The authors have ingeniously devised a regularization loss to ensure the independence and complementarity of these two branches. In the realm of object segmentation, AOT [31] pioneers the use of a long short-term Transformer module, enabling hierarchical feature matching and label propagation. AOT stands out by its capability to concurrently segment multiple objects in a single forward pass, thanks to a purpose-built identification mechanism. Nevertheless, it's worth noting that these methods still rely on point-to-point similarity computations, which can potentially introduce unwanted noise into the predictions.

AGNN [3, 4] achieves video object segmentation by constructing fully connected graphs. AGNN treats each video frame in a video sequence as a graph node. The rich and higher-order inter-frame relationships are mined through iterative information fusion. AGNN adds a gating mechanism to the message aggregation process of the designed graph neural network, which enhances the robustness of the aggregation process. After the graph neural network completes its message passing, the final segmentation result is obtained by a small fully convolutional network (FCN). EGMN [5] enhances AGNN by designing a graph memory network. The learnable read and write controllers of the graph memory network can efficiently update the target appearance representation through meta-learning methods. EGNN samples a fixed number of video frames from the segmented history frames as graph nodes. Nodes features are extracted from the video frames with their masks by a backbone network.

Currently, there is less research related to video object segmentation methods based on graph neural networks. Existing methods treat a video frame as a graph node without fully considering the object structure information within the video frame. Due to their powerful feature learning capabilities, the Transformer methods have been introduced into various fields including video object segmentation and graph representation learning. However, researchers have not yet developed a method that combines graph neural networks and Transformers for video object segmentation. For the first time, we integrate a graph convolutional network and Transformer into a unified framework for video object segmentation.

Specifically, Swin Transformer, a backbone network based on visual Transformer, is utilized. Swin Transformer designs a shifted window scheme with linear computational complexity. Swin Transformer performs self-attention computation within a local window and then achieves cross-window connection through shifted windows. Next, we designed structural feature extraction modules and placed them in different phases of the backbone network. The structural feature extraction module consists of a graph structure construction module and a graph feature enhancement module. The graph structure construction module is implemented based on an agglomerative clustering algorithm (Sect. 7.2.2).

The edge weight matrix of the graph structure output by the graph construction module, and the node features are fed into the graph feature enhancement module,

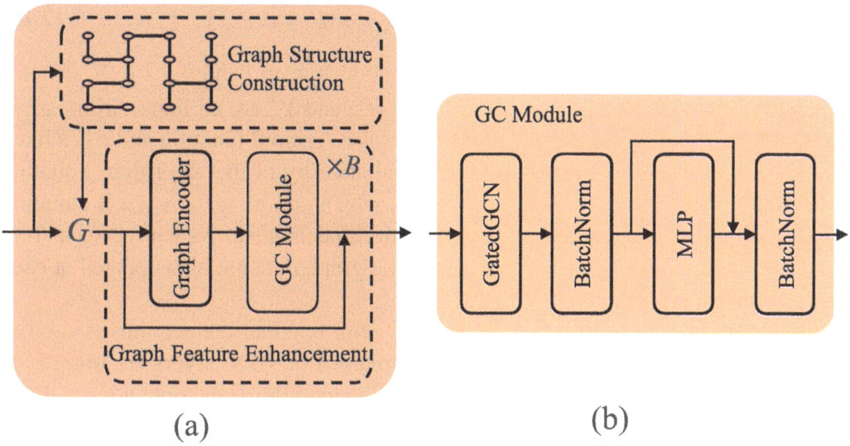

Fig. 7.5 **a** Structural feature extraction module and **b** Graph convolution module (GC module)

as shown in Fig. 7.5. The edge weight matrix of the graph structure and the node features form the graph data. The graph feature enhancement module consists of a graph encoder and B graph convolution module, with residual connections. The graph convolution module consists of a GatedGCN layer, two batch normalization (BN) layers and a multi-layer perceptron (MLP). The MLP consists of two fully connected layers with residual connections between its input and output, as shown in Fig. 7.5.

In the graph encoder, edge weights are encoded as edge features through a fully connected layer. The edge features have the same channel dimension as the channel dimension of the input node features. The graph structure is encoded using a random walk matrix to obtain a structure encoding vector for each graph node. The node features and the structure encoding vector are catenated together after passing through the fully connected layer separately as the updated node features.

The GatedGCN [32] layer is employed in the graph convolution module. GatedGCN is proposed by Bresson and Laurent, which models edge features explicitly. GatedGCN also uses residual connection and achieves better results than methods such as GCN [33] and MoNet [34] on numerous benchmark datasets. The convolution formula of GatedGCN is as follows:

$$h_i^{\ell+1} = h_i^{\ell} + \mathrm{ReLU}\left(\mathrm{BN}\left(U^{\ell} h_i^{\ell} + \sum_{j \in N_i} \hat{e}_{ij}^{\ell} \odot V^{\ell} h_j^{\ell}\right)\right), \tag{7.9}$$

$$\hat{e}_{ij}^{\ell} = \frac{\sigma\left(e_{ij}^{\ell+1}\right)}{\sum_{j' \in N_i} \sigma\left(e_{ij'}^{\ell+1}\right) + \varepsilon}, \tag{7.10}$$

$$e_{ij}^{\ell+1} = e_{ij}^{\ell} + \text{ReLU}\left(\text{BN}\left(A^{\ell}h_i^{\ell} + B^{\ell}h_j^{\ell} + C^{\ell}\hat{e}_{ij}^{\ell}\right)\right), \tag{7.11}$$

where h_i^{ℓ} denotes the node features of the ith node at layer l and h_j^{ℓ}, $j \in \mathcal{N}_i$ denotes its neighbourhood node features. A^{ℓ}, B^{ℓ}, C^{ℓ}, U^{ℓ}, and V^{ℓ} are all linear layers. e_{ij}^{ℓ} is the edge features of nodes i and j at layer l. σ is a sigmoid function and ε is a fixed minimal number used to ensure the numerical stability of the algorithm. Equation (7.10) implements the edge gate mechanism, which can be seen as a soft attention process, similar to the standard sparse attention mechanism. As can be seen from Eqs. (7.10) and (7.11), the edge features of the graph data are also updated at each graph convolution layer.

Similar to Transformer, a two-layer MLP with residual connections is used in the graph convolution module, and batch normalization (BN) layers are inserted to enhance the performance of the graph convolution module. The output node features of the graph feature enhancement module have the same dimension as the input node features.

7.4 Framework and Procedure

In this section, we first outline the overall framework of the proposed method. Then, we elaborate on the designed Object-level Matching (OM) mechanism based on prototype features.

Network Overview. The proposed method consists of four parts: feature extraction, memory construction and updating, feature matching, and decoder, as shown in Fig. 7.6. The feature extraction module consists of the Swin Transformer [26] backbone network and the structural feature extraction module. The memory contains a long-term memory, a short-term memory and a prototype memory. Feature matching includes global matching of the current frame features with the features in the long-term memory, local matching of the current frame features with the features in the short-term memory, and prototype matching of the current frame features with the features in the prototype memory. The feature matching results are used to enhance the current frame features. The current frame features are summed with the enhanced current features obtained by feature matching and then fed into the decoder to obtain the video object segmentation results for the current frame.

The structural feature extraction module is designed and placed at different stages of the feature extraction backbone network to enhance the target structural features. The structural feature extraction module consists of a graph structure generation module and a graph feature extraction module. In order to introduce object-level matching, the proposed method models the target feature distribution by the Gaussian mixture model and uses N Gaussian mean vectors as the prototype features of the target object, which are stored in the prototype memory. Then, matching between the current frame features and the target prototype features are performed and the matching results are utilized to obtain the enhanced current frame features.

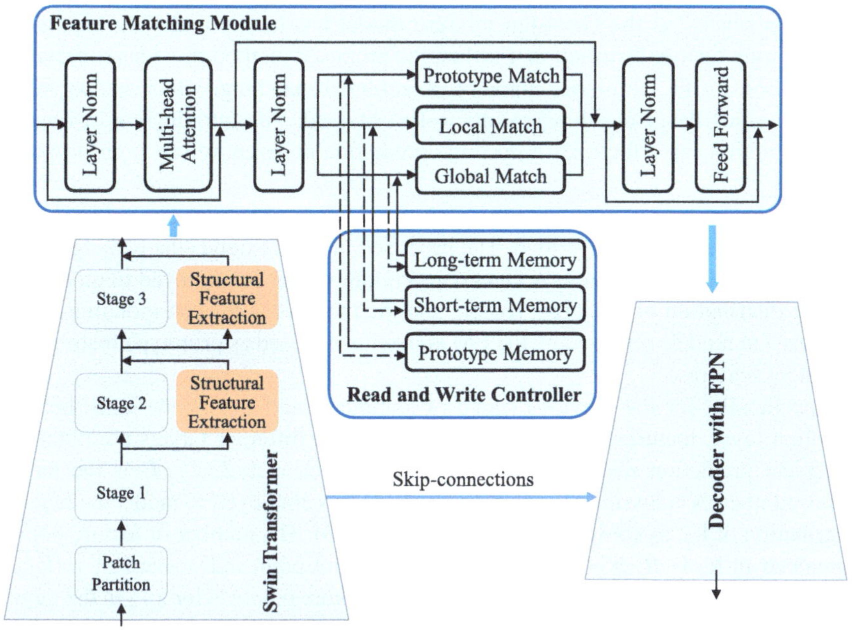

Fig. 7.6 The graph structure output by the graph structure construction module

Object-level matching via prototype features. Recently, researchers use memories to store historical segmentation information and use an attention mechanism to read the segmentation information in the memories, which further enhances the performance of the matching-based video object segmentation method and achieves satisfactory video object segmentation results. However, existing methods usually perform feature matching or compute attention weights based on pixel-level features. This point-to-point similarity calculation does not consider target object-level information and can easily lead to noisy predictions. Yang et al. [35] perform pixel-level matching and design an instance attention mechanism. The authors generated attention vectors by aggregating target region features to guide the segmentation process of the network. However, Yang et al. use global averaging pooling to obtain an aggregated representation of the target. Global average pooling cannot fully utilize target features. The aggregated representation of the target is a single vector, which cannot fully represent the object level information of the target.

Liu et al. [36] use k "word vectors" obtained by the k-means algorithm to represent the target object. Then, the authors calculate the matching scores between the current frame features and all the "word vectors". Inspired by this, the propose method employs N prototype vectors to represent the target object and perform object-level matching, thus fully utilizing the target information. Unlike Liu et al. [36], the proposed method uses a Gaussian mixture model to model the target feature distribution, with N Gaussian mean vectors as the prototype features representing the target object, as the Gaussian mixture model has better modelling capability for the target feature distribution. Further, the proposed method provides supervised information for the similarity maps of object-level matching, which allows more accurate object-level matching results to be obtained. The introduction of object-level matching can effectively reduce the prediction noise caused by point-to-point matching.

First, prototype feature extraction is elaborated. Then, object-level matching based on prototype features is described. The proposed method extends the EM algorithm to video object segmentation. A Gaussian mixture model is employed to model the feature distribution of the target region, and the mean vectors corresponding to the N Gaussian models obtained by the EM algorithm are used as prototype features for the target features.

Specifically, for the previous frame of features $\mathbf{k}$ and $\mathbf{v}$ after the multi-headed attention layer, features $\mathbf{k}_d$ and $\mathbf{v}_d$ corresponding to different targets are obtained using the prediction results of the previous frame. $d = 1, 2, \ldots, D$ is the target index number. A Gaussian Mixture Model (GMM) is employed to model the feature distribution of $\mathbf{k}_d$, as shown in Eqs. (7.12) and (7.13). The number of feature points contained in $\mathbf{k}_d$ is B, $b = 1, 2, \ldots, B$ is the feature point index number. $p\left(\mathbf{k}_{d,b}\right)$ denotes the probability of occurrence of the bth feature point vector $\mathbf{k}_{d,b}$ in the target feature $\mathbf{k}_d$ of the previous frame, obtained by accumulating N Gaussian distributions. $j = 1, 2, \ldots, N$ denotes the index number of the Gaussian model. $p\left(\mathbf{k}_{d,b}|z = j\right)$ denotes the probability value of the feature point vector $\mathbf{k}_{d,b}$ in the jth Gaussian model. z is the hidden variable of membership degree, and Eq. (7.14) further gives the calculation of its posterior probability $p\left(z = j|\mathbf{k}_{d,b}\right)$. M is the dimension size of each feature point vector, and σ is the variance of the Gaussian model, which is a constant. $\mathbf{k}_{d,j}^{o}$ is the class centre of the jth Gaussian model, i.e. a prototype feature. The number of clusters is the number of Gaussian models N. The EM algorithm is employed to find the N clustering centres, i.e. all the prototype features $\mathbf{k}_d^{o}$.

$$p\left(\mathbf{k}_{d,b}\right) = \frac{1}{N}\sum_{j=1}^{N} p\left(\mathbf{k}_{d,b}|z = j\right), \tag{7.12}$$

$$p\left(\mathbf{k}_{d,b}|z = j\right) = \frac{1}{\left(2\pi\sigma^2\right)^{\frac{M}{2}}} \exp\left(-\frac{1}{2\sigma^2}\left\|\mathbf{k}_{d,b} - \mathbf{k}_{d,j}^{o}\right\|^2\right). \tag{7.13}$$

The posterior probability of the hidden variable z can be found according to Bayes' rule to achieve soft clustering of the target features $\mathbf{k}_d$ in the previous frame, as shown in Eq. (7.14). $p\left(z = j\right)$ denotes the probability that the jth Gaussian model occurs in all Gaussian models, which is a uniform distribution i.e. $p\left(z = j\right) = \frac{1}{N}$, $j, l = 1, 2, \ldots, N$ are the index numbers of the Gaussian models and N is the number. By using a negative clustering distance $\left\|\mathbf{k}_{d,b} - \mathbf{k}_{d,l}^{o}\right\|^2$ and a temperature coefficient $2\sigma^2$, Eq. (7.14) can be implemented using the Softmax operation.

$$p\left(z=j|\mathbf{k}_{d,b}\right)=\frac{p\left(\mathbf{k}_{d,b}|z=j\right)p(z=j)}{\sum_{l=1}^{N}p\left(\mathbf{k}_{d,b}|z=l\right)p(z=l)}$$
$$=\frac{\exp\left(-\frac{1}{2\sigma^2}\left\|\mathbf{k}_{d,b}-\mathbf{k}_{d,l}^{o}\right\|^2\right)}{\sum_{l=1}^{N}\exp\left(-\frac{1}{2\sigma^2}\left\|\mathbf{k}_{d,b}-\mathbf{k}_{d,l}^{o}\right\|^2\right)}. \tag{7.14}$$

The value $\mathbf{v}_d^o$ of the prototype feature is calculated using the posterior probability $p\left(z=j|\mathbf{k}_{d,b}\right)$ of z obtained from Eq. (7.14) and the value $\mathbf{v}_d$ of the target feature from the previous frame. The jth feature point vector $\mathbf{v}_{d,j}^o$ in $\mathbf{v}_d^o$ is: $\mathbf{v}_{d,j}^o=\sum_{b=1}^{B}p\left(z=j|\mathbf{k}_{d,b}\right)\mathbf{v}_{d,b}$.

As shown in Fig. 7.6, the long-term memory stores the key-value pairs obtained by self-attention after concatenating the first and periodic frame features, and the short-term memory stores the key-value pairs obtained by self-attention for the previous frame features.

The prototype memory stores the prototype features of the first frame, the periodic frame and the previous frame respectively. Prototype matching is performed between the current frame features and the prototype features of the first frame, the periodic frame and the previous frame, respectively, and the three augmented features obtained are summed as the prototype-based current frame augmented features.

To further constrain the resulting object-level matching map, additional supervised information is introduced. The dimension of the matching map $DN\times HW$, D is the total number of targets including the background, and N is the number of prototype features for each target. The N prototype features of each target in the matching map are summed and softmaxed to obtain the $D\times HW$ matching map, and the $D\times H\times W$ matching map is obtained after shape transformation. The cross entropy of the matching map and the predicted probability map output by the network is calculated as the matching loss.

7.5 Experimental Results and Discussions

In this section, we evaluate the proposed joint graph convolution and Transformer based video object segmentation method on popular datasets. First, the experimental protocol outlines the datasets and evaluation metrics. Then, the implementation details are provided. Subsequently, a comparison is conducted between the proposed method and the state-of-the-art on Youtube-VOS, DAVIS 2016, and DAVIS 2017. Furthermore, qualitative results are presented and analyzed. Finally, a thorough ablation study is performed.

Datasets and metrics for evaluation. In this chapter, we utilize three evaluation datasets: DAVIS 2016 [37], DAVIS 2017 [38], and Youtube-VOS [39].

DAVIS includes both 2016 and 2017 versions. DAVIS 2016 is an extensive high-definition (HD) video dataset, which offers both 480p and 1080p resolutions. Each frame of the videos is annotated, and the dataset consists of 50 videos for training and 30 videos for validation. DAVIS 2016 serves as a dataset for single object video segmentation. Expanding upon DAVIS 2016, DAVIS 2017 introduces additional features. DAVIS 2017 includes annotations for multiple objects and incorporates more video sequences. The training set of DAVIS 2017 encompasses 60 videos, while the validation set and test-dev set contain 30 videos each. Notably, the DAVIS 2017 test-dev set presents a greater challenge as it includes object classes that do not appear in the training set. The DAVSI dataset employs two metrics for evaluation: regional similarity $\mathcal{J}$ and contour accuracy $\mathcal{F}$.

The region similarity coefficient $\mathcal{J}$, also known as Jaccard similarity coefficient, is used to measure the similarity between the algorithm segmentation results and the ground-truth results, such as the number of mis-segmented pixels. Specifically, the Jaccard similarity coefficient $\mathcal{J}$ is obtained by calculating the intersection-over-union between the algorithm's predicted results and the ground-truth annotations, which was first used as an evaluation metric in PASCAL VOC 2008. The Jaccard similarity coefficient is a commonly utilized metric due to its ability to offer intuitive and scale-invariant insights into the number of mis-segmented pixels. For a given video frame, M is used to denote the prediction result of the algorithm and G is used to denote the ground-truth label. The Jaccard similarity coefficient $\mathcal{J}$ is calculated as shown in Eq. (7.15).

$$\mathcal{J} = \frac{|M \cap G|}{|M \cup G|}. \tag{7.15}$$

The contour accuracy $\mathcal{F}$ is used to evaluate the boundary segmentation quality of the segmentation result. The contour of the predicted result M of a video frame is denoted using $c(M)$, and also $c(M)$ defines the spatial extent of the predicted segmentation result M. The segmentation result M is also referred to as the segmentation mask. The precision Pc and recall Rc between the predicted contour $c(M)$ and the ground-truth contour $c(G)$ can then be computed using the bipartite graph matching method. The use of bipartite graph matching enhances the resilience to small contour inaccuracies. The F-measure is used to synthesize the contour precision rate P_c and the contour recall rate R_c, as shown in Eq. (7.16). Considering the efficiency, the evaluation program uses morphological operations to approximate bipartite graph matching.

$$\mathcal{F} = \frac{2P_c R_C}{P_c + R_C}. \tag{7.16}$$

YouTube-VOS stands as a extensive benchmark for video object segmentation, serving as the foundation for subsequent large-scale VOS challenges. Initially introduced in 2018, YouTube-VOS has played a vital role in advancing the field. The training set of YouTube-VOS comprises a remarkable 3,471 videos, spanning across 65 different categories. The validation set of YouTube-VOS is divided into *seen* set and *unseen* set. The *unseen* set encompasses both *seen* and *unseen* categories. The *unseen* set consists of 65 *seen* categories, which were part of the training set, and an additional 26 *unseen* categories. The labels for the validation set are withheld, and the evaluation results are obtained through an evaluation server.

Notably, the YouTube-VOS validation set is available in two versions, each accompanied by its respective evaluation server. Moreover, the validation 2019 split further expands upon the validation 2018 split by introducing more video sequences. The used indicators are $\mathcal{J}_s$ and $\mathcal{F}_s$, $\mathcal{J}_u$ and $\mathcal{F}_u$ and the average of the four indicators is used as the overall score $\mathcal{G}$, as shown in Eq. (7.17).

$$\mathcal{J}\&\mathcal{F} = \frac{\mathcal{J} + \mathcal{F}}{2}. \tag{7.17}$$

Implementation details. The proposed method adopts the architecture of AOT [31] and replaces its backbone network with Swin-B [26], a visual Transformer backbone network. The feature matching process is all done in a modified Transformer module, as shown in Fig. 7.6. The feature matching consists of global matching, local matching and prototype matching, while the enhanced features of the current frame are obtained using the matching results. The long-term memory, short-term memory and prototype memory are constructed for each of the three types of feature matching. The current frame features first go through the layer normalization and multi-head self-attention layer, and then through the layer normalization and feedforward layer after the enhanced features are obtained by feature matching.

Specifically, during global matching, the long-term memory stores $\mathbf{k}_1$ and $\mathbf{v}_1$ of the first frame features. The similarity between the current frame features $\mathbf{k}_c$ with the first frame features $\mathbf{k}_1$ is computed to obtain the global matching map. The global matching map and the first frame feature $\mathbf{v}_1$ are used to obtain the enhanced features of the current frame. The local matching process is similar to the process described above, but the range of matches is limited, with a feature only being matched to features within a specific window size of its location. Prototype matching also uses the matching and enhancement process described above, except that the prototype features are divided into prototype features for the first frame, the periodic frame and the previous frame. The matching and enhancement process is applied to each of these three prototype features, and the enhanced features obtained are summed as the enhanced features based on prototype matching.

Finally, the summation of all matching enhancement features with the current frame features is fed to the decoder. In this case, a dropout layer with a loss rate of 0.1 is used for the summation of all matching enhancement features during training. As with AOT, three modified Transformer modules are stacked in the proposed method for the feature matching process.

The proposed approach incorporates three loss functions to train the network: bootstrap cross-entropy loss and soft IoU loss [40] are utilized between the output segmentation results and the annotation masks, while the prototype matching map contributes to the matching loss. The total loss is computed as the average of these three losses. During the network training phase, not only the Youtube-VOS and DAVIS datasets, as literature [5, 31, 41, 42], but also static image datasets are utilized. The training process consists of two distinct phases. In the initial phase, synthetic videos from the image datasets [43–47] are used for training. In the subsequent stage, network training is performed using randomly sampled video sequences from the video datasets, incorporating video data augmentation techniques as outlined in [48]. Following the methodology described in literature [48], during the training phase, the previous prediction is employed as the mask for the previous frame in each training video, ensuring consistency with the test phase. All videos are downsampled to 480p resolution, and subsequently, 465×465 crop windows are applied to extract video sequences. The chosen length of the video sequence is set to 5 frames.

We employ the AdamW [49] optimizer with a polynomial learning rate for our training process. The training is conducted on two RTX 1080Ti GPUs, utilizing the gradient accumulation technique. In the initial stage of training, we perform 100k training steps on synthetic videos. The learning rate and weight decay for this stage are set to $4 \cdot 1e\text{-}4$ and 0.03, respectively. Moving on to the second stage, where we train on video datasets, we perform 10k training steps for Youtube-VOS and 50k steps for DAVIS. For both Youtube-VOS and DAVIS, the initial learning rate is $2 \cdot 1e\text{-}4$, and the weight decay is set to 0.07. Throughout the training process, we use a batch size of 16. We gradually decay the learning rate to $2 \cdot 1e\text{-}5$. To prevent overfitting, we apply an additional factor of 0.1 to the initial learning rate of the encoder. Additionally, we adopt the Exponential Moving Average (EMA) technique.

Comparison with state-of-the-arts. In this section, we conduct a comprehensive comparison between our proposed method and recent state-of-the-art approaches on DAVIS 2016 [37], DAVIS 2017 [38], and Youtube-VOS [39] datasets. To indicate whether synthetic data is used for network training, we include flags in the "S" column of the tables. Additionally, we use flags in the "OL" column to specify whether online fine-tuning is employed. In the tables, we present the backbone networks used by all the methods. Specifically, MobileNetV2 [50] is denoted as MN2, ResNet-50 [51] as R50, ResNet-101 [51] as R101, and WideResNet-38 [52] as WR38. Moreover, we represent DeepLabv2 [53] and DeepLabv3+ [54] as DL2 and DL3+, respectively. VGG-m is a backbone network proposed by Chatfield et al. [55]. For comprehensive comparison, we also include the performance of our proposed method when using the R50 backbone network, indicated as SFEP-R (Table 7.1).

Compared to DAVIS 2016, DAVIS 2017 is used for multi-object segmentation and adds more video sequences. The DAVIS validation set and test-dev set contain 59 and 89 objects, respectively. The DAVIS test-dev set is more challenging as it contains object classes that do not appear in the training set. As presented in Table VIII, our proposed method SFEP-R achieves an impressive overall score of 86.1%, surpassing the baseline method and showcasing the effectiveness of our approach. Notably, SFEP, which utilizes Swin-B as the backbone network, outperforms SFEP-R in terms of performance. Furthermore, our proposed method demonstrates superior performance compared to most other methods, including previously advanced techniques such as STCN* [65] and HMMN* [42]. The STCN [65] effectively reduces computational complexity by employing a lightweight module for frame feature and segmentation mask encoding, eliminating the need for redundant feature encoding for each object. On the other hand, the HMMN [42] introduces a kernel-guided memory matching module and a top-k guided memory matching module to enhance

Table 7.1 Quantitative comparison with state-of-the-art methods on the DAVIS 2017 [38] validation set. S indicates using synthetic data for pre-training. OL denotes online learning. t/s remarks second per frame. † indicates that the time is inferred from the single-object time. The best three results are boldfaced, italicized, and underlined, respectively. * indicates using both DAVIS and Youtube-VOS for training

Method	Backbone	S	OL	$\mathcal{J}\&\mathcal{F}$	$\mathcal{J}$	$\mathcal{F}$	t/s
OSVOS[CVPR17] [56]	VGG16	×	✓	60.3	56.6	63.9	17.7†
OnAVOS[CVPRW17] [57]	WR38	×	✓	67.9	64.5	71.2	26†
RGMP[CVPR18] [41]	R50	✓	×	66.7	64.8	68.6	0.28†
OSVOS-S[PAMI19] [58]	VGG16	×	✓	68.0	64.7	71.3	8.9†
FEELVOS[CVPR19] [19]	DL3+	×	×	71.5	69.1	74.0	0.51
SiamMask[CVPR19] [59]	R50	×	×	56.4	54.3	58.5	0.06†
RANet[ICCV19] [20]	R101	✓	×	65.7	63.2	68.2	0.06†
STM[ICCV19] [21]	R101	✓	×	71.6	69.2	74.0	0.32†
MGCRN[PAMI20] [60]	R101	×	×	68.1	64.8	71.4	1.44†
KMN[ECCV20] [61]	R50	✓	×	76.0	74.2	77.8	0.24†
CFBI[ECCV20] [48]	R101	×	×	74.9	72.1	77.7	0.17
LWTL[ECCV20] [62]	R50	×	×	74.3	72.2	76.3	0.4†
LCM[CVPR21] [22]	R50	✓	×	75.2	73.1	77.2	0.23†
RMNet[CVPR21] [23]	R50	✓	×	75.0	72.8	77.2	0.28†
SST[CVPR21] [29]	R101	×	×	78.4	75.4	81.4	–
JOINT[ICCV21] [30]	R50	×	×	78.6	76.0	81.2	0.25
HMMN[ICCV21] [42]	R50	✓	×	80.4	77.7	83.1	0.20†
STM*[ICCV19] [21]	R50	✓	×	81.8	79.2	84.3	0.32†
SAT*[CVPR20] [63]	R50	×	×	72.3	68.6	76.0	*0.05*†
EGMN*[ECCV20] [5]	R50	✓	×	82.8	80.2	85.2	0.4†
KMN*[ECCV20] [61]	R50	✓	×	82.8	80.0	85.6	0.24†
CFBI*[ECCV20] [48]	R101	×	×	81.9	79.3	84.5	0.17
LWTL*[ECCV20] [62]	R50	×	×	81.6	79.1	84.1	0.4†
LCM*[CVPR21] [22]	R50	✓	×	83.5	80.5	86.5	0.23†
RMNet*[CVPR21] [23]	R50	✓	×	83.5	81.0	86.0	0.28†
GIEL*[CVPR21] [24]	R50	×	×	82.7	80.2	85.3	–
SST*[CVPR21] [29]	R101	×	×	82.5	79.9	85.1	–
JOINT*[ICCV21] [30]	R50	×	×	83.5	80.8	86.2	0.25
DMNAOA*[ICCV21] [64]	R101	✓	×	84.5	81.4	87.5	0.16
HMMN*[ICCV21] [42]	R50	✓	×	84.7	81.9	87.5	0.20†
AOT*[NIPS21] [31]	MN2	✓	×	83.0	80.3	85.7	*0.05*
STCN*[NIPS21] [65]	R50	✓	×	85.4	82.2	88.6	*0.05*
CFBI+*[PAMI22] [35]	R101	×	×	82.9	80.1	85.7	0.17

(continued)

Table 7.1 (continued)

Method	Backbone	S	OL	$\mathcal{J}\&\mathcal{F}$	$\mathcal{J}$	$\mathcal{F}$	t/s
ASRF*[TIP22] [66]	DL3+	×	×	83.2	80.3	86.1	–
MaskVOS*[TIP22] [67]	R50	✓	×	83.3	80.0	86.7	–
AOML*[TIP22] [68]	R50	✓	×	83.2	80.3	86.1	0.26
SITVOS*[AAAI22] [69]	R50	✓	×	83.5	80.4	86.5	0.8
PCVOS*[CVPR22] [70]	R50	✓	×	86.1	83.0	89.2	–
DeAOT*[NIPS22] [71]	Swin-B	✓	×	86.2	83.1	89.2	0.06
XMem*[ECCV22] [72]	R50	✓	×	**88.2**	**85.4**	**91.0**	**0.04**
ISVOS*[CVPR23] [73]	R50	✓	×	*87.1*	*83.7*	*90.5*	–
SFEP-R*	R50	✓	×	86.1	83.2	88.9	0.10
SFEP*	Swin-B	✓	×	86.5	83.3	89.7	0.15

STM [21]. These modules are strategically designed to strengthen spatio-temporal coherence and leverage multi-scale information, respectively.

However, STCN and HMMN still adopt a point-to-point matching strategy and are susceptible to interference from similar background objects or regions. The proposed method introduces object-level matching information based on prototype features, which can effectively mitigate noise prediction. The proposed method adopts an improved Transformer module to achieve feature matching. Compared with a simple one-time computed matching map, the improved Transformer module contains a self-attention mechanism and can be stacked with multiple modules, which can more fully exploit intra- and inter-frame target feature similarity information.

XMem [72] and ISVOS [73] obtain higher performance than SFEP. XMem designes three types of memory: sensor memory, working memory and long-term memory, inspired by the Atkinson-Shiffrin model. ISVOS improves dense memory-based matching by integrating an instance segmentation branch. The introduction of the instance segmentation branch gives the network the ability to understand instances. Further, the results of all methods on the test-dev set of DAVIS 2017 are shown in Table 7.2. The proposed method still outperforms the baseline method and most other methods with an overall score of 82.3%. The results on the DAVIS 2017 test-dev set demonstrate the effectiveness of the proposed method SFEP.

The validation set of DAVIS 2016 contains 20 video sequences for single object evaluation. As shown in Table 7.3, the proposed method SFEP obtains an overall performance of 92.6%, outperforming almost all methods. The DAVIS 2016 validation set contains fewer video sequences and relatively simple video scenes, so the evaluation results on this dataset tend to be saturated. XMem [72] and ISVOS [73] have achieved higher performance. As mentioned above, XMem stores frequently used features from working memory into long-term memory via a memory potentiation algorithm. ISVOS introduces the knowledge of objects learned in the instance segmentation branch to dense memory-based matching.

Table 7.2 Quantitative comparison with state-of-the-art methods on the DAVIS 2017 [38] test-dev set. S indicates using synthetic data for pre-training. OL denotes online learning. The best three results are boldfaced, italicized, and underlined, respectively. * indicates using both DAVIS and Youtube-VOS for training

Method	Backbone	S	OL	$\mathcal{J}\&\mathcal{F}$	$\mathcal{J}$	$\mathcal{F}$
OnAVOS[CVPRW17] [57]	WR38	×	✓	56.5	53.4	59.6
PReMVOS[ACCV18] [74]	R101	×	✓	71.6	67.5	75.7
RGMP[CVPR18] [41]	R50	✓	×	52.9	51.3	54.4
OSVOS-S[PAMI19] [58]	VGG16	×	✓	57.5	–	–
FEELVOS*[CVPR19] [19]	DL3+	×	×	57.8	55.2	60.5
RANet[ICCV19] [20]	R101	✓	×	55.3	53.4	57.2
STM*[ICCV19] [21]	R50	✓	×	72.2	69.3	75.2
KMN*[ECCV20] [61]	R50	✓	×	77.2	74.1	80.3
CFBI*[ECCV20] [48]	R101	×	×	75.0	71.4	78.7
LCM*[CVPR21] [22]	R50	✓	×	78.1	74.4	81.8
RMNet*[CVPR21] [23]	R50	✓	×	75.0	71.9	78.1
GIEL*[CVPR21] [24]	R50	×	×	75.2	72.0	78.3
DMNAOA*[ICCV21] [64]	R101	✓	×	78.3	74.8	81.7
HMMN*[ICCV21] [42]	R50	✓	×	78.6	74.7	82.5
AOT*[NIPS21] [31]	MN2	✓	×	78.4	74.8	82.1
STCN*[NIPS21] [65]	R50	✓	×	76.1	72.7	79.6
CFBI+*[PAMI22] [35]	R101	×	×	75.6	71.6	79.6
DeAOT*[NIPS22] [71]	Swin-B	✓	×	*82.8*	78.9	**86.7**
XMem*[ECCV22] [72]	R50	✓	×	**83.1**	**79.7**	*86.4*
ISVOS*[CVPR23] [73]	R50	✓	×	*82.8*	*79.3*	86.2
SFEP-R*	R50	✓	×	80.7	77.2	84.2
SFEP*	Swin-B	✓	×	82.3	78.7	85.9

The Youtube-VOS validation set is available in two versions, 2018 and 2019. The validation set for the 2018 version contains 474 videos. As shown in Table 7.4, the proposed method outperforms the baseline method with an overall score of 85.5%. Besides, the proposed method SFEP outperforms the majority of other methods, except XMmem [72], ISVOS [73] and DeAOT [71]. DeAOT [71] boosts AOT [31] by designing dual-branch propagation, which decouples object-agnostic features from object-specific features. The 2019 version contains 504 video sequences. The proposed method outperforms the baseline method and other existing methods, as indicated in Table 7.5. XMem [72] and ISVOS [73] achieve higher performance. XMem stores frequently used features from working memory into long-term memory through a memory potentiation algorithm. ISVOS improves dense memory-based matching by integrating an instance segmentation branch, as mentioned above. Experimental results on both versions of the validation set demonstrate the effectiveness of SFEP.

Table 7.3 Quantitative comparison with state-of-the-art methods on the DAVIS 2016 [37] validation set. S indicates using synthetic data for pre-training. OL denotes online learning. The best three results are boldfaced, italicized, and underlined, respectively. * indicates using both DAVIS and Youtube-VOS for training

Method	Backbone	S	OL	$\mathcal{J}\&\mathcal{F}$	$\mathcal{J}$	$\mathcal{F}$
MaskTrack[CVPR17] [75]	DL2	×	✓	77.6	79.7	75.4
OSVOS[CVPR17] [56]	VGG16	×	✓	80.2	79.8	80.6
OnAVOS[CVPRW17] [57]	WR38	×	✓	85.5	86.1	84.9
PReMVOS[ACCV18] [74]	R101	×	✓	86.8	84.9	88.6
MoNet[CVPR18] [76]	DL2	×	✓	84.8	84.7	84.8
RGMP[CVPR18] [41]	R50	✓	×	81.8	81.5	82.0
OSVOS-S[PAMI19] [58]	VGG16	×	✓	86.5	85.6	87.5
STCNN[CVPR19] [77]	R101	×	✓	83.8	83.8	83.8
FEELVOS[CVPR19] [19]	DL3+	×	×	81.7	81.1	82.2
SiamMask[CVPR19] [59]	R50	×	×	69.8	71.7	82.2
AGAME[CVPR19] [78]	R101	✓	×	82.1	82.2	82.0
RANet+[CVPR19] [20]	R101	✓	✓	87.1	86.6	87.6
STM*[ICCV19] [21]	R50	✓	×	89.3	88.7	89.9
MGCRN[PAMI20] [60]	R101	×	×	85.1	84.4	85.7
SAT*[CVPR20] [63]	R50	×	×	83.1	82.6	83.6
KMN*[ECCV20] [61]	R50	✓	×	90.5	89.5	91.5
CFBI*[ECCV20] [48]	R101	×	×	89.4	88.3	90.5
LCM*[CVPR21] [22]	R50	✓	×	90.7	89.9	91.4
RMNet*[CVPR21] [23]	R50	✓	×	88.8	88.9	88.7
HMMN*[ICCV21] [42]	R50	✓	×	90.8	89.6	92.0
AOT*[NIPS21] [31]	MN2	✓	×	91.0	89.7	92.3
STCN*[NIPS21] [65]	R50	✓	×	91.6	90.8	92.5
GraphVOS[PAMI22] [79]	VGG-m	×	×	90.2	89.2	91.2
CFBI+*[PAMI22] [35]	R101	×	×	89.9	88.7	91.1
ASRF*[TIP22] [66]	DL3+	×	×	90.9	90.1	91.7
MaskVOS*[TIP22] [67]	R50	✓	×	90.0	89.2	90.8
AOML*[TIP22] [68]	R50	✓	×	92.1	91.2	93.0
SITVOS*[AAAI22] [69]	R50	✓	×	90.5	89.5	91.4
PCVOS*[CVPR22] [70]	R50	✓	×	91.9	90.8	93.0
DeAOT*[NIPS22] [71]	Swin-B	✓	×	**92.9**	91.1	**94.7**
XMem*[ECCV22] [72]	R50	✓	×	*92.7*	**92.0**	93.5
ISVOS*[CVPR23] [73]	R50	✓	×	92.6	91.5	*93.7*
SFEP-R*	R50	✓	×	91.7	91.1	92.3
SFEP*	Swin-B	✓	×	92.6	*91.6*	93.5

Table 7.4 Quantitative comparison with state-of-the-art methods on Youtube-VOS [39] *Validation 2018 Split*. S indicates using synthetic data for pre-training. OL denotes online learning. t/s remarks second per frame. † indicates that the time is inferred from the single-object time. The best three results are boldfaced, italicized, and underlined, respectively

Method	Backbone	S	OL	$\mathcal{G}$	$\mathcal{J}_{seen}$	$\mathcal{F}_{seen}$	$\mathcal{J}_{unseen}$	$\mathcal{F}_{unseen}$	t/s
OSVOS[CVPR17] [56]	VGG16	×	✓	58.8	59.8	60.5	54.2	60.7	–
OnAVOS[CVPRW17] [57]	WR38	×	✓	55.2	60.1	62.7	46.1	51.4	–
PReMVOS[ACCV18] [74]	R101	×	✓	66.9	71.4	75.9	56.5	63.7	5.88
RGMP[CVPR18] [41]	R50	✓	×	53.8	59.5	–	45.2	–	–
S2S[ECCV18] [80]	VGG16	×	×	57.6	66.7	65.5	48.2	50.3	–
SiamMask[CVPR19] [59]	R50	×	×	52.8	60.2	58.2	45.1	47.7	–
RVOS[CVPR19] [81]	R101	×	×	56.8	63.6	67.2	45.5	51.0	–
AGAME[CVPR19] [78]	R101	✓	×	66.0	66.9	–	61.2	–	–
STM[ICCV19] [21]	R50	✓	×	79.4	79.7	84.2	72.8	80.9	–
SAT[CVPR20] [63]	R50	×	×	63.6	67.1	70.2	55.3	61.7	–
EGMN[ECCV20] [5]	R50	✓	×	80.2	80.7	85.1	74.0	80.9	–
KMN[ECCV20] [61]	R50	✓	×	81.4	81.4	85.6	75.3	83.3	–
CFBI[ECCV20] [48]	R101	×	×	81.4	81.1	85.8	75.3	83.4	0.29
LWTL[ECCV20] [62]	R50	×	×	81.5	80.4	84.9	76.4	84.4	–
LCM[CVPR21] [22]	R50	✓	×	82.0	82.2	86.7	75.7	83.4	–
RMNet[CVPR21] [23]	R50	✓	×	81.5	82.1	85.7	75.7	82.4	–
GIEL[CVPR21] [24]	R50	×	×	80.6	80.7	85.0	75.0	81.9	–
SST[CVPR21] [29]	R101	×	×	81.7	81.2	–	76.0	–	–
JOINT[ICCV21] [30]	R50	×	×	83.1	81.5	85.9	78.7	86.5	–
DMNAOA[ICCV21] [64]	R101	✓	×	82.7	82.6	87.0	76.7	84.8	–
HMMN[ICCV21] [42]	R50	✓	×	82.6	82.1	87.0	76.8	84.6	–
AOT[NIPS21] [31]	MN2	✓	×	83.7	82.5	87.5	77.9	86.7	*0.06*
STCN[NIPS21] [65]	R50	✓	×	83.0	81.9	86.5	77.9	85.7	–
GraphVOS[PAMI22] [79]	VGG-m	×	×	73.4	74.4	78.6	66.5	73.9	–
CFBI+[PAMI22] [35]	R101	×	×	82.8	81.8	86.6	77.1	85.6	0.25
ASRF[TIP22] [66]	DL3+	×	×	81.9	81.0	85.8	76.3	84.3	–
MaskVOS[TIP22] [67]	R50	✓	×	81.5	80.5	85.2	76.0	84.3	–
AOML[TIP22] [68]	R50	✓	×	83.8	82.4	87.3	78.6	86.8	–
SITVOS[AAAI22] [69]	R50	✓	×	81.3	79.9	84.3	76.4	84.4	–
PCVOS[CVPR22] [70]	R50	✓	×	84.6	83.0	88.0	79.6	87.9	–
DeAOT[NIPS22] [71]	Swin-B	✓	×	86.2	**85.6**	**90.6**	80.0	88.4	0.08
XMem[ECCV22] [72]	R50	✓	×	**86.7**	85.3	89.9	**81.7**	**89.9**	**0.04**
ISVOS[CVPR23] [73]	R50	✓	×	*86.3*	*85.5*	*90.2*	*80.5*	*88.8*	–
SFEP-R	R50	✓	×	85.1	84.3	88.9	79.6	87.6	0.12
SFEP	Swin-B	✓	×	85.5	84.8	89.7	79.4	87.9	0.19

Table 7.5 Quantitative comparison with state-of-the-art methods on Youtube-VOS [39] *Validation 2019 Split*. S indicates using synthetic data for pre-training. OL denotes online learning. t/s remarks second per frame. † indicates that the time is inferred from the single-object time. The best three results are boldfaced, italicized, and underlined, respectively

Method	Backbone	S	OL	$\mathcal{G}$	$\mathcal{J}_{seen}$	$\mathcal{F}_{seen}$	$\mathcal{J}_{unseen}$	$\mathcal{F}_{unseen}$	t/s
STM[ICCV19] [21]	R50	✓	×	79.2	79.6	83.6	73.0	80.6	–
CFBI[ECCV20] [48]	R50	×	×	81.0	80.6	85.1	75.2	83.0	0.29
LWTL[ECCV20] [62]	R50	×	×	81.0	79.6	83.8	76.4	84.2	–
SST[CVPR21] [29]	R101	×	×	81.8	80.9	–	76.6	–	
JOINT[ICCV21] [30]	R50	×	×	82.8	80.8	84.8	79.0	86.6	–
HMMN[ICCV21] [42]	R50	✓	×	82.5	81.7	86.1	77.3	85.0	–
AOT[NIPS21] [31]	MN2	✓	×	83.6	82.2	86.9	78.3	86.9	*0.06*
STCN[NIPS21] [65]	R50	✓	×	82.7	81.1	85.4	78.2	85.9	–
CFBI+[PAMI22] [35]	R101	×	×	82.6	81.7	86.2	77.1	85.2	0.25
AOML[TIP22] [68]	R50	✓	×	83.7	82.1	86.6	79.1	87.1	–
PCVOS[CVPR22] [70]	R50	✓	×	84.6	82.6	87.3	80.0	88.3	–
DeAOT[NIPS22] [71]	Swin-B	✓	×	*86.1*	**85.3**	**90.2**	80.4	88.6	0.08
XMem[ECCV22] [72]	R50	✓	×	**86.4**	84.9	89.2	**81.8**	**89.8**	**0.04**
ISVOS[CVPR23] [73]	R50	✓	×	*86.1*	*85.2*	*89.7*	*80.7*	*88.9*	–
SFEP-R	R50	✓	×	84.9	83.7	88.2	80.2	87.6	0.12
SFEP	Swin-B	✓	×	85.4	84.2	88.9	80.4	88.2	0.19

Figure 7.7 shows the segmentation results of the proposed method for some video sequences in the DAVIS 2017 dataset. The first frame of the video and its target object ground-truth mask are shown on the left, while the segmentation results for subsequent video frames are shown on the right. It can be seen that the proposed method SFEP can distinguish between similar objects, such as not being influenced by the second camel that appears in subsequent frames of the first video.

The proposed method can effectively deal with scenes in which people interact with objects, such as a person and a four-wheeled vehicle in the third video and a person and a box being carried in the fourth video. The proposed method can accurately segment branching targets such as the rope referenced in the surfing in the second video and the handrail of the quad bike in the fourth video. The proposed method can handle complex backgrounds, such as the more complex background interference in the fourth video.

Ablation studies. We further analyze the key components Structural Feature Extraction (SFE), Object-level Matching (OM), and hyper-parameters. The results of the ablation experiments with the structural feature enhancement module and the object-level matching mechanism on the validation set from DAVIS 2017 are shown in Table 7.6.

Naive baseline. The architecture of AOT [31] is adopted by the baseline approach. The original backbone network is replaced with Swin-B [26], which is a backbone network based on visual Transformer. AOT is one of the representative methods for

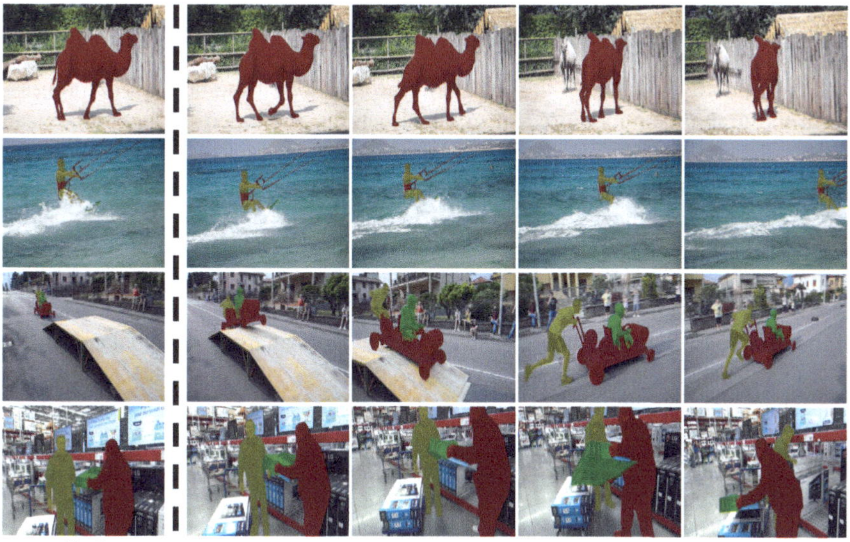

Fig. 7.7 Quantitative results of the proposed method on four challenging sequences

Table 7.6 Ablation experiments on the DAVIS 2017 validation set. The structural feature enhancement module and the object-level matching mechanism are components of the proposed method

Version	Structural feature enhancement	Object-level matching	$\mathcal{J}\&\mathcal{F}$
Naive			85.1
Sophisticate	✓		85.8
Sophisticate		✓	86.0
SFEP	✓	✓	86.5

matching-based video object segmentation, which introduces an identity embedding mechanism to segment multiple video objects simultaneously in a single forward computation. AOT designes a long short-term Transformer module for modelling intra-frame similarity and long short-term inter-frame similarity in videos. The proposed Structural Feature Enhancement (SFE) module, Object-level Matching (OM) mechenism, are then added to the baseline method.

SFE. The structural feature enhancement module is added to increase the backbone network's ability to extract target structural features, resulting in more accurate target segmentation. Compared to the baseline approach, the addition of only the SFE module resulted in an 0.7% performance improvement, as shown in Table 7.6. The structural feature enhancement module consists of a graph structure generation module and a graph feature extraction module. Figure 7.4 shows the graph structure output from the graph structure generation module, which is connected within the target and background regions, corresponding to different sub-minimum spanning trees. The generated graph structure is disconnected at the boundaries of the different regions, separating the different sub-minimum spanning trees. The generated graph structure effectively captures the structural information of the target. The processing of the graph data composed of graph structures and video frame features can enhance the ability of the backbone network to extract features of the target structure by utilizing graph convolutional networks.

OM. Object-level matching mechanisms are added to reduce the prediction noise caused by point-to-point matching. As shown in Table 7.6, the proposed method obtains a 0.9% improvement over the baseline method when only the OM module is added. The Gaussian mixture model is employed to model the target feature distribution and the EM algorithm is used to solve the Gaussian mixture model consisting of N Gaussians to obtain the target prototype features. The prototype features are then used to match with the current frame features to achieve object-level matching to increase the robustness of the matching process. As shown in Fig. 7.7, the third video has interaction and occlusion between the human and the four-wheeled vehicle, and the fourth video has more complex background interference, and the proposed method achieves promising segmentation results in these challenging scenes. When both SFE module and OM mechanism are added, the proposed method obtains a 1.4% improvement compared to the baseline method, indicating that the two modules can be used together.

We further conducts full experiments on the hyper-parameters in the SFE module. The effect of the constant c, which controls the size of the region in the graph generation module, on the performance of the method is first analyzed. Table 7.7 shows the experimental results for different values of c. The performance scores of the proposed method fluctuate less when the value of c is taken to vary in the interval [50,100]. We set c to 80 to obtain a higher performance score. The effect of the number of graph convolution blocks on the performance of the method is then analysed. As shown in Table 7.8, the method performance rises and then slightly decreases as the number of graph convolution blocks increases. The number of graph convolution blocks is set to 4 to obtain a high performance score and saves computational overhead.

We then analyze the impact of the location of the graph feature extraction modules in the backbone network on the performance of the method. The Swin series backbone

Table 7.7 Results of ablation experiments for c. c controls the size of the subregion in the graph generation module

c	50	60	70	80	90	100
$\mathcal{J}\&\mathcal{F}$	86.2	86.3	86.4	86.5	86.4	86.2

Table 7.8 Results of ablation experiments for B. B denotes the number of graph convolution blocks in the graph feature extraction module

B	1	2	3	4	5	6
$\mathcal{J}\&\mathcal{F}$	86.1	86.2	86.4	86.5	86.5	86.4

Table 7.9 Ablation experiments on the DAVIS 2017 validation set. The structural feature enhancement modules are placed at different stages of the backbone network

Stage	2	3	2, 3
$\mathcal{J}\&\mathcal{F}$	86.1	86.3	86.5

Table 7.10 Ablation experiments on the DAVIS 2017 validation set. Different numbers of prototypes are used to represent each target region features

Proto. Number	1	8	16	24	32
$\mathcal{J}\&\mathcal{F}$	86.2	86.4	86.5	86.4	86.3

Table 7.11 Results of ablation experiments for the number of EM iterations

Iteration Number	1	2	4	6	8
$\mathcal{J}\&\mathcal{F}$	86.2	86.4	86.4	86.5	86.3

network can be divided into five stages from 0 to 4. Stage 0 performs patch splitting and feature mapping. The remaining stages consist of patch merging and different numbers of Swin blocks. As shown in Table 7.9, the performance of the method when place in stage 3 only is higher than that when place in stage 2 only. We place the graph feature extraction module in both stages 2 and 3 to obtain higher performance scores.

In addition, we further analyze the impact of the components and hyper-parameters in the OM mechanism on the performance of the method. The impact of the number of prototypes representing each target region features on the performance of the method is first analysed. As shown in Table 7.10, as the number of prototypes increases, the performance of the proposed method first rises and then falls. When the number of prototypes is greater than 16, the effect of object-level matching based on prototypes is weakened. We then analyze the effect of the number of iterations in the EM algorithm on the performance of the method. Table 7.11 shows the results for different numbers of iterations. As the number of iterations increases from 1 to 6, the performance of the method gradually increases. When the number of iterations is greater than or equal to 6, the method performance is saturated. The number of iterations is therefore set to 6.

We further replace GMM with k-means and the results are shown in Table 7.12. The method performance when using only GMM is higher than when using only k-means. This indicates that GMM has a better ability to model the target region feature distribution compared to k-means. The ablation results for matching loss are shown in Table 7.12, indicating that matching loss has a positive effect on both k-means and GMM. We use both GMM and matching loss to obtain higher performance scores.

This chapter presents a structural feature enhancement module and a prototype feature extraction algorithm for video object segmentation. The structural feature

Table 7.12 Ablation experiments on the DAVIS 2017 validation set. Results when using Gaussian Mixture Models (GMM) or k-means to obtain prototypes. The result when matching loss is used or not

Version	GMM	k-means	Matching loss	$\mathcal{J}\&\mathcal{F}$
C1		✓		86.1
C2	✓			86.3
C3	✓		✓	86.5
SFEP		✓	✓	86.3

enhancement module is developed to capture and enhance the target structural features. The target region has certain structural shape features, such as straight lines and corners contained in targets like poles and cars. The Graph Convolutional Networks have received a lot of attention from researchers for their powerful structural feature extraction capabilities. In this chapter, we design a graph structure generation algorithm and use a graph convolutional network to process the obtained graph data, which can effectively enhance the target structure features to obtain more robust video frame features. Further, we use multiple prototype features as an aggregated representation of the target region features to fully utilize the target feature information and thus obtain more accurate object-level matching maps.

Deep supervision is used by many methods [82, 83] to improve network performance and accelerate convergence. Therefore, we use the prediction probabilities of the output layer to provide additional supervised information for the object-level matching map. At the application level, the proposed structural feature enhancement module and prototype feature extraction algorithm can be extended to related fields such as object tracking and action recognition. In future work, we will further adapt and streamline the network architecture of the proposed method in terms of both memory footprint and inference speed.

References

1. Sun, J., Gao, L., Shen, X., Liu, S., Liang, R., Du, S., Liu, S.: Separated graph neural networks for recommendation systems. IEEE Transactions on Industrial Informatics 19(1), 382–393 (2023)
2. Malik, J., Shimada, S., Elhayek, A., Ali, S.A., Theobalt, C., Golyanik, V., Stricker, D.: Handvoxnet++: 3d hand shape and pose estimation using voxel-based neural networks. IEEE Transactions on Pattern Analysis and Machine Intelligence 44(12), 8962–8974 (2022)
3. Wang, W., Lu, X., Shen, J., Crandall, J.D., Shao, L.: Zero-shot video object segmentation via attentive graph neural networks. In: Proceedings of the IEEE International Conference on Computer Vision, pp. 9236–9245 (2019)
4. Lu, X., Wang, W., Shen, J., Crandall, D.J., Van Gool, L.: Segmenting objects from relational visual data. IEEE Transactions on Pattern Analysis and Machine Intelligence 44(11), 7885–7897 (2022)

5. Lu, X., Wang, W., Danelljan, M., Zhou, T., Shen, J., Gool, V.L.: Video object segmentation with episodic graph memory networks. In: Proceedings of the European Conference Computer Vision, pp. 661–679 (2020)
6. Song, L., Li, Y., Jiang, Z., Li, Z., Zhang, X., Sun, H., Sun, J., Zheng, N.: Rethinking learnable tree filter for generic feature transform. In: Advances in Neural Information Processing Systems, pp. 3991–4002 (2020)
7. Hou, Q., Zhang, L., Cheng, M.M., Feng, J.: Strip pooling: Rethinking spatial pooling for scene parsing. In: Proceedings of the IEEE Conference on Computer Vision and Pattern Recognition, pp. 4003–4012 (2020)
8. Gori, M., Monfardini, G., Scarselli, F.: A new model for learning in graph domains. In: Proceedings. 2005 IEEE International Joint Conference on Neural Networks, 2005., vol. 2, pp. 729–734 vol. 2 (2005)
9. Vaswani, A., Shazeer, N., Parmar, N., Uszkoreit, J., Jones, L., Gomez, N.A., Kaiser, L., Polosukhin, I.: Attention is all you need. In: Advances in Neural Information Processing Systems, vol. 30, pp. 5998–6008 (2017)
10. Voigtlaender, P., Chai, Y., Schroff, F., Adam, H., Leibe, B., Chen, L.C.: Non-local neural networks. In: Proceedings of the IEEE Conference on Computer Vision and Pattern Recognition, pp. 9473–9482 (2018)
11. Yang, Z., Wei, Y., Yang, Y.: Hierarchical question-image co-attention for visual question answering. In: Advances in Neural Information Processing Systems (2016)
12. Felzenszwalb, P.F., Huttenlocher, D.P.: Efficient graph-based image segmentation. Int. J. Computer Vision 59, 167–181 (2004)
13. Yang, Z., Wei, Y., Yang, Y.: Rethinking graph transformers with spectral attention. In: Advances in Neural Information Processing Systems (2021)
14. Malik, J., Shimada, S., Elhayek, A., Ali, S.A., Theobalt, C., Golyanik, V., Stricker, D.: Spectral grouping using the nystrom method. IEEE Transactions on Pattern Analysis and Machine Intelligence 26(2), 214–225 (2004)
15. Yang, Z., Wei, Y., Yang, Y.: Distance encoding: Design provably more powerful neural networks for graph representation learning. In: Advances in Neural Information Processing Systems (2020)
16. Yang, Z., Wei, Y., Yang, Y.: Graph neural networks with learnable structural and positional representations. In: Advances in Neural Information Processing Systems (2022)
17. Chen, Y., Pont-Tuset, J., Montes, A., Gool, L.: Blazingly fast video object segmentation with pixel-wise metric learning. In: Proceedings of the IEEE Conf Computer Vision and Pattern Recognition (CVPR), pp. 1189–1198 (2018)
18. Hu, Y.T., Huang, J.B., Schwing, A.G.: VideoMatch: Matching Based Video Object Segmentation. In: Proceedings of the European Conference Computer Vision, pp. 56–73 (2018)
19. Voigtlaender, P., Chai, Y., Schroff, F., Adam, H., Leibe, B., Chen, L.C.: FEELVOS: Fast end-to-end embedding learning for video object segmentation. In: Proceedings of the IEEE Conference on Computer Vision and Pattern Recognition, pp. 9473–9482 (2019)
20. Wang, Z., Xu, J., Liu, L., Zhu, F., Shao, L.: RANet: Ranking attention network for fast video object segmentation. In: Proceedings of the IEEE International Conference on Computer Vision, pp. 3977–3986 (2019)
21. Oh, W.S., Lee, J.Y., Xu, N., Kim, J.S.: Video object segmentation using space-time memory networks. In: Proceedings of the IEEE International Conference on Computer Vision, pp. 9225–9234 (2019)
22. Li, H., Peng, Z., Bang, Z., Pan, P., Yinghui, X., Rong, J.: Learning position and target consistency for memory-based video object segmentation. In: Proceedings of the IEEE Conference on Computer Vision and Pattern Recognition, pp. 4144–4154 (2021)
23. Xie, H., Yao, H., Zhou, S., Zhang, S., Sun, W.: Efficient regional memory network for video object segmentation. In: Proceedings of the IEEE Conference on Computer Vision and Pattern Recognition, pp. 1286–1295 (2021)
24. Ge, W., Lu, X., Shen, J.: Video object segmentation using global and instance embedding learning. In: Proceedings of the IEEE Conference on Computer Vision and Pattern Recognition, pp. 16836–16845 (2021)

25. Dosovitskiy, A., Beyer, L., Kolesnikov, A., Weissenborn, D., Zhai, X., Unterthiner, T., Dehghani, M., Minderer, M., Heigold, G., Gelly, S., Uszkoreit, J., Houlsby, N.: An image is worth 16×16 words: Transformers for image recognition at scale. arXiv:2010.11929 (2020)
26. Liu, Z., Lin, Y., Cao, Y., Hu, H., Wei, Y., Zhang, Z., Lin, S., Guo, B.: Swin transformer: Hierarchical vision transformer using shifted windows. In: Proceedings of the IEEE International Conference on Computer Vision, pp. 9992–10002 (2021)
27. Carion, N., Massa, F., Synnaeve, G., Usunier, N., Kirillov, A., Zagoruyko, S.: End-to-end object detection with transformers. In: Proceedings of the European Conference Computer Vision, pp. 213–229 (2020)
28. Dong, B., Zeng, F., Wang, T., Zhang, X., Wei, Y.: SOLQ: Segmenting objects by learning queries. In: Advances in Neural Information Processing Systems, vol. 34, pp. 21898–21909 (2021)
29. Duke, B., Ahmed, A., Wolf, C., Aarabi, P., Taylor, W.G.: Sstvos: Sparse spatiotemporal transformers for video object segmentation. In: Proceedings of the IEEE Conference on Computer Vision and Pattern Recognition, pp. 5912–5921 (2021)
30. Mao, Y., Wang, N., Zhou, W., Li, H.: Joint inductive and transductive learning for video object segmentation. In: Proceedings of the IEEE International Conference on Computer Vision, pp. 9650–9659 (2021)
31. Yang, Z., Wei, Y., Yang, Y.: Associating objects with transformers for video object segmentation. In: Advances in Neural Information Processing Systems, vol. 34, pp. 2491–2502 (2021)
32. Bresson, X., Laurent, T.: Residual gated graph convnets. arXiv:1711.07553 (2017)
33. Kipf, T.N., Welling, M.: Semi-supervised classification with graph convolutional networks. In: Proceedings of the Inter. Conf. Learning Representations (ICLR) (2017)
34. Monti, F., Boscaini, D., Masci, J., Rodola, E., Svoboda, J., Bronstein, M.M.: Geometric deep learning on graphs and manifolds using mixture model cnns. In: Proceedings of the IEEE Conference on Computer Vision and Pattern Recognition, pp. 5115–5124 (2017)
35. Yang, Z., Wei, Y., Yang, Y.: Collaborative video object segmentation by multi-scale foreground-background integration. IEEE Transactions on Pattern Analysis and Machine Intelligence 44(9), 4701–4712 (2022)
36. Liu, Y., Liu, L., Zhang, H., Rezatofighi, H., Yan, Q., Reid, I.: Meta learning with differentiable closed-form solver for fast video object segmentation. In: Proceedings of the Int. Conf. on Intell. Robots and Sys. (IROS), pp. 8439–8446 (2020)
37. Perazzi, F., Pont-Tuset, J., Mcwilliams, B., Gool, L.V., Gross, M., Sorkine-Hornung, A.: A benchmark dataset and evaluation methodology for video object segmentation. In: Proceedings of the IEEE Conference on Computer Vision and Pattern Recognition, pp. 724–732 (2016)
38. Pont-Tuset, J., Perazzi, F., Caelles, S., Arbeláez, P., Sorkine-Hornung, A., Gool, L.V.: The 2017 DAVIS challenge on video object segmentation. arXiv:1704.00675 (2017)
39. Xu, N., Yang, L., Fan, Y., Yue, D., Liang, Y., Yang, J., Huang, T.: YouTube-VOS: A large-scale video object segmentation benchmark. arXiv:1809.03327 (2018)
40. Nowozin, S.: Optimal decisions from probabilistic models: the intersection-over-union case. In: Proceedings of the IEEE Conference on Computer Vision and Pattern Recognition, pp. 548–555 (2014)
41. Oh, S., Lee, J.Y., Sunkavalli, K., Kim, S.: Fast video object segmentation by reference-guided mask propagation. In: Proceedings of the IEEE Conference on Computer Vision and Pattern Recognition, pp. 7376–7385 (2018)
42. Seong, H., Oh, S.W., Lee, J.Y., Lee, S., Lee, S., Kim, E.: Hierarchical memory matching network for video object segmentation. In: Proceedings of the IEEE International Conference on Computer Vision, pp. 12869–12878 (2021)
43. Everingham, M., Gool, L., Williams, K.C., Winn, J., Zisserman, A.: The pascal visual object classes (voc) challenge. Int. J. Computer Vision 88(2), 303–338 (2010)
44. Lin, T.Y., Maire, M., Belongie, S., Hays, J., Perona, P., Ramanan, D., Dollár, P., Zitnick, L.C.: Microsoft coco: Common objects in context. In: Proceedings of the European Conference Computer Vision, pp. 740–755 (2014)

45. Cheng, M.M., Zhang, G.X., Mitra, N.J., Huang, X., Hu, S.M.: Global contrast based salient region detection. IEEE Transactions on Pattern Analysis and Machine Intelligence 37(3), 569–582 (2014)
46. Shi, J., Yan, Q., Xu, L., Jia, J.: Hierarchical image saliency detection on extended cssd. IEEE Transactions on Pattern Analysis and Machine Intelligence 38(4), 717–729 (2015)
47. Hariharan, B., Arbeláez, P., Bourdev, L., Maji, S., Malik, J.: Semantic contours from inverse detectors. In: Proceedings of the IEEE International Conference on Computer Vision, pp. 991–998 (2011)
48. Yang, Z., Wei, Y., Yang, Y.: Collaborative video object segmentation by foreground-background integration. In: Proceedings of the European Conference Computer Vision, pp. 332–348 (2020)
49. Loshchilov, I., Hutter, F.: Decoupled weight decay regularization. In: Proceedings of the Inter. Conf. Learning Representations (ICLR) (2019)
50. Sandler, M., Howard, A., Zhu, M., Zhmoginov, A., Chen, L.C.: MobileNetv2: Inverted residuals and linear bottlenecks. In: Proceedings of the IEEE Conference on Computer Vision and Pattern Recognition, pp. 4510–4520 (2018)
51. He, K., Zhang, X., Ren, S., Sun, J.: Deep residual learning for image recognition. In: Proceedings of the IEEE Conference on Computer Vision and Pattern Recognition, pp. 770–778 (2016)
52. Wu, Z., Shen, C., Hengel, A.v.d.: Wider or deeper: Revisiting the ResNet model for visual recognition. Pattern Recognition 90, 119 – 133 (2019)
53. Chen, L.C., Papandreou, G., Kokkinos, I., Murphy, K., Yuille, A.L.: DeepLab: Semantic image segmentation with deep convolutional nets, atrous convolution, and fully connected crfs. IEEE Transactions on Pattern Analysis and Machine Intelligence 40(4), 834–848 (2018)
54. Chen, L.C., Zhu, Y., Papandreou, G., Schroff, F., Adam, H.: Collaborative video object segmentation by foreground-background integration. In: Proceedings of the European Conference Computer Vision, pp. 833–851 (2018)
55. Chatfield, K., Simonyan, K., Vedaldi, A., Zisserman, A.: Return of the devil in the details: Delving deep into convolutional nets. In: Proceedings of the British Machine Vision Conference (BMVC) (2014)
56. Caelles, S., Maninis, K.K., Pont-Tuset, J., Leal-Taixé, L., Cremers, D., Gool, L.V.: One-shot video object segmentation. In: Proceedings of the IEEE Conference on Computer Vision and Pattern Recognition, pp. 5320–5329 (2017)
57. Voigtlaender, P., Leibe, B.: Online adaptation of convolutional neural networks for the 2017 davis challenge on video object segmentation. The 2017 DAVIS Challenge on Video Object Segmentation - CVPR Workshops (2017)
58. Maninis, K.K., Caelles, S., Chen, Y., Pont-Tuset, J., Leal-Taixe, L., Cremers, D., Van Gool, L.: Video object segmentation without temporal information. IEEE Transactions on Pattern Analysis and Machine Intelligence 41, 1515–1530 (2019)
59. Wang, Q., Zhang, L., Bertinetto, L., Hu, W., Torr, H.S.P.: Fast online object tracking and segmentation: A unifying approach. In: Proceedings of the IEEE Conference on Computer Vision and Pattern Recognition, pp. 1328–1338 (2019)
60. Hu, P., Wang, G., Kong, X., Kuen, J., Tan, Y.: Motion-guided cascaded refinement network for video object segmentation. IEEE Transactions on Pattern Analysis and Machine Intelligence 42, 1957–1967 (2020)
61. Seong, H., Hyun, J., Kim, E.: Kernelized memory network for video object segmentation. In: Proceedings of the European Conference Computer Vision, pp. 629–645 (2020)
62. Goutam, B., Järemo, F.L., Martin, D., Andreas, R., Michael, F., Luc, G.V., Radu, T.: Learning what to learn for video object segmentation. In: Proceedings of the European Conference on Computer Vision, pp. 777–794 (2020)
63. Xi, C., Zuoxin, L., Ye, Y., Gang, Y., Jianxin, S., Donglian, Q.: State-aware tracker for real-time video object segmentation. In: Proceedings of the IEEE Conference on Computer Vision and Pattern Recognition, pp. 9381–9390 (2020)
64. Liang, S., Shen, X., Huang, J., Hua, X.S.: Video object segmentation with dynamic memory networks and adaptive object alignment. In: Proceedings of the IEEE International Conference on Computer Vision, pp. 8045–8054 (2021)

65. Cheng, H.K., Tai, Y.W., Tang, C.K.: Rethinking space-time networks with improved memory coverage for efficient video object segmentation. In: Advances in Neural Information Processing Systems, vol. 34, pp. 11781–11794 (2021)
66. Hong, L., Zhang, W., Chen, L., Zhang, W., Fan, J.: Adaptive selection of reference frames for video object segmentation. IEEE Transactions on Image Processing 31, 1057–1071 (2022)
67. Wang, M., Mei, J., Liu, L., Tian, G., Liu, Y., Pan, Z.: Delving deeper into mask utilization in video object segmentation. IEEE Transactions on Image Processing 31, 6255–6266 (2022)
68. Guo, P., Zhang, W., Li, X., Zhang, W.: Adaptive online mutual learning bi-decoders for video object segmentation. IEEE Transactions on Image Processing 31, 7063–7077 (2022)
69. Lan, M., Zhang, J., He, F., Zhang, L.: Siamese network with interactive transformer for video object segmentation. In: Proceedings of the AAAI Conference on Artificial Intelligence, pp. 1228–1236 (2022)
70. Park, K., Woo, S., Oh, S.W., Kweon, I.S., Lee, J.Y.: Per-clip video object segmentation. In: Proceedings of the IEEE Conference on Computer Vision and Pattern Recognition, pp. 1352–1361 (2022)
71. Yang, Z., Yang, Y.: Decoupling features in hierarchical propagation for video object segmentation. In: Advances in Neural Information Processing Systems, vol. 35, pp. 36324–36336 (2022)
72. Cheng, H.K., Schwing, A.G.: XMem: Long-Term video object segmentation with an atkinson-shiffrin memory model. In: Proceedings of the European Conference Computer Vision, pp. 640–658 (2022)
73. Wang, J., Chen, D., Wu, Z., Luo, C., Tang, C., Dai, X., Zhao, Y., Xie, Y., Yuan, L., Jiang, Y.G.: Look before you match: Instance understanding matters in video object segmentation. In: Proceedings of the IEEE Conference on Computer Vision and Pattern Recognition (2023)
74. Luiten, J., Voigtlaender, P., Leibe, B.: PReMVOS: Proposal-generation, refinement and merging for video object segmentation. In: Proceedings of the Asian Conference Computer Vision (ACCV), pp. 565–580 (2018)
75. Perazzi, F., Khoreva, A., Benenson, R., Schiele, B., Sorkine-Hornung, A.: Learning video object segmentation from static images. In: Proceedings of the IEEE Conference on Computer Vision and Pattern Recognition, pp. 3491–3500 (2017)
76. Xiao, H., Feng, J., Lin, G., Liu, Y., Zhang, M.: MoNet: Deep motion exploitation for video object segmentation. In: Proceedings of the IEEE Conference on Computer Vision and Pattern Recognition, pp. 1140–1148 (2018)
77. Xu, K., Wen, L., Li, G., Bo, L., Huang, Q.: Spatiotemporal cnn for video object segmentation. In: Proceedings of the IEEE Conference on Computer Vision and Pattern Recognition, pp. 1379–1388 (2019)
78. Johnander, J., Danelljan, M., Brissman, E., Khan, F.S., Felsberg, M.: A generative appearance model for end-to-end video object segmentation. In: Proceedings of the IEEE Conference on Computer Vision and Pattern Recognition, pp. 8945–8954 (2019)
79. Giraldo, J.H., Javed, S., Bouwmans, T.: Graph moving object segmentation. IEEE Transactions on Pattern Analysis and Machine Intelligence 44(5), 2485–2503 (2022)
80. Xu, N., Yang, L., Fan, Y., Yang, J., Yue, D., Liang, Y., Price, B., Cohen, S., Huang, T.: YouTube-VOS: Sequence-to-sequence video object segmentation. In: Proceedings of the European Conference Computer Vision, pp. 603–619 (2018)
81. Ventura, C., Bellver, M., Girbau, A., Salvador, A., Marqués, F., i Nieto, X.G.: RVOS: End-to-end recurrent network for video object segmentation. In: Proceedings of the IEEE Conference on Computer Vision and Pattern Recognition, pp. 5272–5281 (2019)
82. Zhao, J.X., Liu, J.J., Fan, D.P., Cao, Y., Yang, J., Cheng, M.M.: Egnet: Edge guidance network for salient object detection. In: Proceedings of the IEEE International Conference on Computer Vision, pp. 8779–8788 (2019)
83. Acuna, D., Kar, A., Fidler, S.: Devil is in the edges: Learning semantic boundaries from noisy annotations. In: Proceedings of the IEEE Conference on Computer Vision and Pattern Recognition, pp. 11075–11083 (2019)

Chapter 8
Conclusion and Future Work

This book aims to introduce readers to the fundamental concepts of graph neural network learning and its application in scene understanding. Graph neural networks have gained significant attention in recent years due to their effectiveness in handling structured data like graphs and networks. These networks are especially useful for tasks like scene understanding, where relationships and interactions between elements play a crucial role.

8.1 Summary

In this book, we first introduce the history and foundational concepts of graph representation learning, and the history and frontiers of graph neural networks. In the section on graph representation learning, we introduce the concepts and properties of graphs, and then lead to spectral methods on graphs and graph embedding methods. Graph neural networks further advance the development of graph representation learning. We focus on the history of the development of vanilla graph neural networks, graph convolutional networks, and graph attention networks. Besides, we give cutting-edge advances in graph neural networks, including graph generative models, dynamic graph neural networks, and graph neural architecture search.

Then, we focus on scene understanding, which is a kind of computer vision task to get category of place, structure of object, attribute of event and other semantic information from the scene. Firstly, according to scene data type, we categorize scene understanding into image-based understanding, video-based understanding and 3D based understanding. And each kind of scene understanding can be divided into many small tasks (L1, L2, and L3) according to task target. Further, in order to better

W. Liu et al., *Graph Neural Network Methods and Applications in Scene Understanding*,
https://doi.org/10.1007/978-981-97-9933-6_8

illustrate the scene understanding task, we show the taxonomy of scene understanding, i.e., scene representation, scene analysis, weak semantic scene understanding and strong semantic scene understanding. Scene understanding include a variety of visual tasks, and in this book we focus on three tasks, i.e., scene parsing, human parsing and video object segmentation. Finally, we conclude the challenges of scene understanding from internal factors and external factors.

Further, we detail the basics of graph neural network. We begin by outlining the general framework of graph neural networks, which primarily includes feature transformation layers, activation layers, and pooling layers. Next, we describe the typical operations and representative methods of graph neural networks. For graph neural network operations, we cover spectral graph convolution, spatial graph convolution, graph attention, and graph pooling. Regarding graph neural network architectures, we present notable methods for vanilla graph neural networks, graph convolutional networks, and graph attention networks, providing readers with a comprehensive understanding of methods for various types of graph neural networks. Finally, we summarize the training methods for graph neural networks.

Scene understanding is a crucial aspect of computer vision and artificial intelligence. It also covers many computer vision tasks. This book mainly focuses on three tasks, i.e., scene parsing, human parsing and video object segmentation. Graph neural networks exhibit excellent performance when dealing with non-Euclidean data. Therefore, this book aims to address the important problems of the three tasks using GNN.

Scene parsing requires the algorithm model to realize pixel-level scene object recognition and segmentation. This book introduces the scene parsing method with Self-supervised feature Fusion-based Graph Convolution Network (SFGCN). Firstly, the hierarchical graph construction is used to construct scene graph feature data in this study, and then the spatial graph convolution operation based on K-level neighbor is used to construct SFGCN model to achieve global feature learning. In order to convert original image data into graph data, a novel hierarchical graph construction method has been proposed, which stacks multi-granularity grids of images (as nodes) and connects them according to their semantic labels. Graph construction projects the original scene image into topological space to obtain scene graph data, and its essence is to transform RGB image data into graph structure data. Then, to solve the problem that local semantic labels are difficult to parse, a graph attention module based on self-supervised feature fusion is proposed. By combining spectral graph convolution and attention mechanism, self-supervised information is added to model weight training to enhance local feature learning. For the feature learning of objects in different scales, on one hand, a Self-supervised Feature fusion-based Attention (SFA) mechanism has been proposed to enhance feature learning for small objects according to local sub-graph structures. On the other hand, an improved spatial graph convolution based on K-level neighbor of nodes, namely K-Conv, has been designed to enlarge the receptive field for feature learning of graph nodes. K-Conv operation mainly focuses on the global feature learning of scene graph data. Finally, in order to preserve the intra-domain consistency of multi-domain scene images, this study proposes a scene domain adaptive model-based graph pooling module to improve the intra-domain inconsistency of scene images caused by their imaging differences. Due to different imaging conditions resulting in large intra-class difference, a Scene Domain Adaptive Pooling (SDAP) mechanism has been added into SFGCN model to maintain the intra-domain consistency of multi-domain scene images. Experimental results on six scene benchmarks show that the proposed method can effectively implement scene parsing and outperforms other advanced methods.

Human parsing promotes the use of image data for visual understanding of the human body at an amazing speed. This book focuses on the graph neural network for human parsing and dynamic graph neural network for human parsing. In graph neural network for human parsing, this book introduces an end-to-end graph neural network based hierarchical human parsing. Further, to solve the problem of high cost of human parsing, this book proposes to use the weakly supervised method for human parsing. In dynamic graph neural network for human parsing, this book makes full use of the natural topology of human body and proposes a dynamic graph neural network for human parsing.

In this book, we introduce an end-to-end graph neural network based hierarchical human parsing, named Context Prior based Semantic-Spatial Graph Network (CP-SSGNet). CP-SSGNet consists of three modules, Semantic Constraint Module (SCM), Spatial Perceiving Module (SPM), and Intra-class Attention Module (IAM). SCM aims to capture contextual information in a global view and reduce errors related to semantic structure. To further improve the quality of local features, CP-SSGNet introduce the SPM. SPM plays a crucial role by encoding the prior knowledge of spatial consistency within the context. IAM utilizes the spatial graph with strong–weak connections for intra-class features aggregation to distinguish the inter-class features clearly. However, high performance human parsing methods need pixel-level annotation for supervised learning. The labor-intensive burden has limited its widespread application in downstream tasks. This book introduce a graph neural network based weakly supervised human parsing method. First, utilizing both box-level labels and point-level labels, this book design a region-semantic based unsupervised segmentation strategy to generate the initial pixel-level pseudo labels. Furthermore, to settle the noisy labels and label scarcity problem of the initial pseudo labels, we propose an iterative training model Dynamic Hierarchical Semantic Graph Neural Network (DHSGNN) to learn more representative features and obtain reliable fine-grained parsing results.

The choice of a fixed graph is no longer suitable when dealing with diverse data in human parsing tasks. Hence, this book proposes to generate an dynamic graph based on multi-label for human parsing. The dynamic graph is achieved by the proposed Graph Node Adaptation (GNA) and Graph Connection Adaptation (GCA). Multilabel learning is a basic task of computer vision. Here, we apply the estimating multilabel classification scores to generate the adaptive matrix, and then use the adaptive matrix to achieve GNA and GCA. Specially, in order to better construct the dynamic graph, we propose a 0–1 polarization function to further optimize this adaptive matrix. Based on the adaptive matrix, GNA successfully removes the features of nodes that may include noise, and GCA studies a adaptive adjacency matrix to achieve a more effective propagation manner. By combining GNA and GCA, the fixed graph can be optimized into a data-driven dynamic graph for each individual training sample. Further, considering the hierarchical structure of human body, we effectively encode the hierarchical semantic structural prior knowledge into the graph representation for human parsing. By generating the dynamic hierarchical semantic graph in a dense-to-coarse manner, we naturally capture and utilize the structural information. We define this process as Dynamic Topology-aware Pooling (DTP) strategy.

Video object segmentation is an active research area in the field of computer vision, which has made great strides in recent years. However, similar objects or backgrounds may interfere with accurate segmentation, leading to noisy predictions. Meanwhile, some targets such as poles and ropes cannot be segmented completely. To capture the structural shape information of the target, we first propose a graph structure generation algorithm based on agglomerative clustering. As to graph construction for video data, we employ the agglomerative clustering algorithm to generate a series of sub-minimum spanning trees. The sub-minimum spanning trees form a graph structure that is disconnected at object boundaries and connected within object regions, thus effectively capturing the target structural information. Based on the graph construction, we employ a graph convolutional network to process the graph data consisting of the graph structure and video frame features to further enhance the structural features of the target. We design structural feature extraction modules and place them in different phases of the backbone network. Swin Transformer is used as the backbone network. In order to reduce the prediction noise caused by the interference of similar objects, we design a prototype matching mechanism. In order to introduce object-level matching, the proposed method extends the EM algorithm to video object segmentation. First, we construct a prototype feature extraction algorithm based on Gaussian mixture model to represent the whole target with a small number of prototype features. Then, we perform prototype feature matching to enhance the robustness of the feature matching process. Prototype matching is performed between the current frame features and the prototype features of the first frame, the periodic frame and the previous frame, respectively. The three obtained augmented features are summed as the prototype-based current frame augmented features. In addition, we add object-level matching loss to the total loss function to introduce more supervised information for network training.

8.2 Future Work

As various scene understanding technologies become increasingly mature, scene understanding is being applied in multiple fields to assist with labor-intensive tasks and improve productivity. Therefore, scene understanding is also facing new challenges. On one hand, there is a need to research more precise and efficient scene understanding technologies to provide reliable technical support for various intelligent analysis and decision-making. On the other hand, with the diversification of application scenarios, there are significant differences between data in different scenarios. Additionally, complex graph relationships exist between the data. The challenge is how to utilize these data characteristics to design effective graph models and enhance the recognition performance in specific scene understanding tasks.

Scene Parsing. Scene parsing has always been a challenging image processing problem in computer vision, and it is also a key issue in geometric understanding in computer graphics. Scene parsing task aims to recognize and analyze the object and environment of the scene through the computer. In this book, we study this problem from the perspective of feature fusion, and the proposed method has achieved certain performance improvement in scene parsing. However, there are still many problems and challenges in this field that have not been completely solved, and more in-depth research is needed.

Scene parsing is a pixel-level recognition task, which needs to realize scene object recognition and contour segmentation at the same time. Besides, it needs to make comprehensive use of scene category, scene attribute and scene semantic information. Due to the high complexity of the scene parsing task, improving the pixel-level parsing accuracy is still the focus of research. The in-depth research of deep learning has further promoted the rapid development of scene parsing tasks. At the same time, it also promotes the accuracy of scene parsing tasks to be constantly breakthrough. However, most of the existing deep learning-based scene parsing models are strongly-supervised models, which over-rely on high-cost pixel-level annotations. Therefore, how to reduce the need for manual annotation of scene semantic models based on deep learning, and how to study scene parsing algorithms based on weak supervision and unsupervised learning, so that they can use massive data for intelligent learning, is the focus of future research. Building smart cities is the main development direction at present, and scene parsing can provide sufficient information in autonomous driving tasks. Hence, it plays an important role in smart cities. In this case, it is not only necessary to improve the accuracy of scene parsing task, but also put forward higher requirements for the real-time performance of scene parsing. Therefore, the high-speed and real-time of scene parsing model has become a key point to transform the theoretical model of scene semantic parsing into practical application. In the follow-up study, this issue will be investigated.

Human Parsing. With the development of graph neural network technology, the performance of human parsing model has also been greatly improved. However, human body data and human posture are rich and changeable. Dynamic graphs can adapt to changes in the data or input conditions. This adaptability is particularly valuable in tasks where the relationships between nodes are not fixed and may change over time. Dynamic graphs can be computationally efficient because they allow for the selective update of graph structure or information propagation. And this can allow the model to learn more comprehensive and rich features. Further, occlusion is a common condition in human body images, and it is also a difficult challenge for human parsing. Dynamic graphs can adapt to occluded regions by adjusting the graph structure or propagating information selectively. Hence, it is worth studying to explore a more effective method of dynamic graph construction for human body.

The weakly supervised based human parsing model can effectively alleviate the substantial manpower and resources required by pixel-level annotation. Furthermore, by leveraging the natural topological structure of the human body, the effective combination of weakly supervised learning and graph neural networks can reduce the dependence on pixel-level supervised information, enable accurate extraction of human semantic features, thus enhance the performance accuracy of the human body parsing model. In the future, exploring how to make use of more types of weak supervision information for human parsing, such as box-level information, image-level information, and scribble-level information, and construct reasonable graph according to various types of weak supervision information is a very worthwhile task. Compared with pixel-level labels required for human parsing, the key point annotation of pose estimation is easier to obtain. Therefore, it is noteworthy to design the mapping relationship between human skeleton graph and constructed graph of human parsing based on graph neural network in topological space. Based on this, we can achieve a high performance weakly supervised human model.

Video Object Segmentation. Current semi-supervised video object segmentation algorithms have made significant progress, and can achieve better results in scenes such as background clutter and dynamic background. However, the length of each video in existing video object segmentation datasets is short, such as the large-scale Youtube-VOS dataset with an average video duration between three and six seconds, and the performance of existing segmentation methods for video object segmentation under minute-level long videos has not been effectively evaluated. The target disappearance is a challenging scenario that often occurs in long videos. It can be specifically categorized into two scenarios, one in which the target disappears and then does not reappear, and the other in which the target disappears and then reappears. In the early studies of video object segmentation, the video scenes in the dataset are relatively simple, and the segmentation methods are more likely to deal with the situation where the target disappears and then no longer appears, and cannot effectively deal with the situation where the target disappears and then reappears.

Referring video object segmentation means that a short text description is given in the first frame of the video to specify the object to be segmented, and then that text description is utilized to segment the video target specified in subsequent frames of the video. The text description is a new form of providing supervisory information for video object segmentation. In recent years, researchers proposed a audio-visual object segmentation method that uses audio to provide supervised information to specify video objects for video object segmentation, where the algorithm needs to segment video targets that emit sounds in audio. Further, it would be more intelligent and convenient for practical applications if a model supports supervisory information from multiple modalities to accomplish video object segmentation, providing multiple choices to the user. The current multimodal related technologies have made breakthroughs in the fields of text image retrieval, artificial intelligence painting, and so on. Driven by multimodal pre-training large models, artificial intelligence technology, etc., multimodal video object segmentation methods will be the future development trend. Currently, referring video object segmentation still has more deficiencies, audio video object segmentation is still in the initial development stage, and there is no model that supports three modal supervised information of video frame mask, text description and audio for video object segmentation research. Therefore, it would be a worthwhile direction to explore how to utilize graph neural networks to promote the development of reference video object segmentation, audio-visual object segmentation, and multimodal video object segmentation that support multimodal supervisory information.

Image-based scene understanding is a fundamental task which deals with RGB image captured by general digital visual sensor. Image-based scene understanding can be classified into general scene, human and text. The subtasks of general scene can be scene classification, scene attribute recognition, scene object detection, and scene graph generation. The subtasks of human can be human/pedestrian attribute recognition. The subtasks of text can be scene text segmentation, scene text recognition and scene text detection.

- Scene Classification. The challenges of scene classification include large intra-class diversity, high inter-class similarity, and large differences in scene scales. In terms of diversity within classes, the challenge comes mainly from the large differences in features that occur within the same semantic class. Ground objects often vary in style, shape, size, and distribution, making it difficult to properly classify scene images. The challenge of interclass similarity is mainly due to the overlap of the same objects between different scene classes or high semantic scene classes. The huge difference of scene scale is also a challenge that cannot be ignored in aerial image scene classification. In satellite imaging, sensors work in orbit at different altitudes, from a few hundred kilometers to more than ten thousand kilometers, and the imaging height varies greatly. Even the same scene has a huge scale difference at different imaging heights.
- Scene Attribute Recognition. Scene attribute recognition is to infer multiple attribute labels of a scene image on the basis of scene representation, so as to have a deeper semantic understanding of the scene. Therefore, attribute recognition is a multi-label classification problem. Although humans are very skilled in perceiving natural scenes and understanding advanced scene structures visually, computers need more complex algorithm models and more image visual features, such as shape, edge, texture and color, to effectively integrate the information of a scene image from multiple angles and capture the interaction between various scene objects. The multi-attribute recognition results of the scene are obtained.
- Scene Object Detection. Object detection has become very mature at the technical level, and a large number of classic and efficient detection algorithms have emerged, such as yolo series, faster RCNN, SSD, etc., but the problem of small target detection is still the focus and difficulty of academic and industrial research. The main difficulties are as follows. (1) Small target coverage image area is small, effective features for target detection are few, and common detectors are not sensitive to small targets; (2) Small targets are scattered in various areas of the image, may appear in corners, may be blocked with large targets, and small targets are easily affected by image noise; (3) AP and mAP indexes commonly used in target detection are not applicable to small targets, and small deviations of small targets will lead to big changes in IoU; (4) Small target detection data sets are few, most

of them are data sets of simple scenes, and there is a lack of small target detection data sets in complex scenes.

- Scene Graph Generation. A scene graph is a structural representation that represents objects in a picture as nodes and their relationships as edges. It is an informative holistic approach to scene understanding that Bridges the huge gap between the visual and natural language domains. While great progress has been made in generating scene diagrams from a single image (static scene diagram generation), the task of generating scene diagrams from video (dynamic scene diagram generation) is new and more challenging. The most popular method of static scene graph generation is based on object detectors and then inferring their relationship types as well as their object classes. However, objects are not necessarily consistent in every frame of the video sequence, and the relationship between any two objects may change due to their motion, with dynamic characteristics. In this case, time dependence plays a role, so the static scene graph generation method cannot be directly applied to dynamic scene graph generation.
- Human/Pedestrian Attribute Recognition. Human attribute recognition is to predict a set of attributes from a predefined attribute list to describe the characteristics of a person in the line. For example, given a graduation photo of a college student, the attributes that identify that person might include short hair, glasses, college dress, black pants, black shoes, blue bicycle, male, and so on. Pedestrian attribute recognition mining is the high-level semantic information of pedestrians, which is different from the low-level features, and is relatively robust to the change of perspective and imaging conditions. The earliest pedestrian attribute recognition was done by manually extracting features and training classifiers separately for each different attribute. With the development of CNN, people begin to try to place all attributes in the same network for multi-task training, share network parameters, and find that multi-task training can bring better results. At present, the basic method of pedestrian attribute recognition is to throw the whole picture into the same CNN network and output multiple labels representing attributes for classification. Based on this basic method, the latest work mainly focuses on how to identify attributes of different granularity and different scale, how to extract context information in the scene to assist the identification of attributes, and how to extract the correlation information between different attributes. Different granularity attributes are extracted by establishing branches of different network levels respectively, and then the features extracted from different branches are spliced as the final features. By establishing the LSTM network structure, the context can be introduced into the subsequent feature extraction process. The correlation information between different attributes is extracted by multi-attribute joint training or by using LSTM network to save the identification information of the previous attribute.
- Scene Text Recognition, Detection and Segmentation. Segmentation based recognition algorithm is an important branch of natural scene text recognition algorithm, which usually includes three steps: image preprocessing, single character segmentation and single character recognition. Segmentation based natural scene text recognition algorithm usually needs to locate the location of each character

contained in the input text image, identify each character through a single character identifier, and then combine all the characters into a string sequence to obtain the final recognition result. The search time of the algorithm is related to the size of the dictionary. With the increase of the size of the dictionary to be matched, the search space of the algorithm is also greatly increased. Therefore, the natural scene text recognition algorithm which relies on dictionary matching to obtain the best recognition result is difficult to be directly applied to the actual application. In order to solve the above difficulties, natural scene text recognition algorithms without dictionary constraints have been widely explored. Using a large-scale dictionary as a high-level statistical language model, this book overcomes the disadvantage of requiring a specific dictionary to match the recognition results, and realizes the natural scene text recognition without dictionary constraints. An early natural scene text recognition system takes the score and position of a single character as the system input, uses a graphical model to find the best matching text corresponding to the recognition result from a specific dictionary, and uses it as the final recognition result.

Video-based scene understanding task is used to analyze continuous image frames, and each frame has a unique timestamp. It needs to extract and analyze information from successive frames to infer various aspects of the scene's content, structure, and dynamics. Videos are captured using cameras or other recording devices which can generate a sequence of frames over time. These frames collectively represent the visual information of the scene. Specially, the video data generally requires preprocessing to enhance quality, stabilize motion and remove noise. In this book, it is classified as general scene and human.

- Video scene parsing is a subtask of general scene. Its objective is to automatically assign predefined semantic labels to pixels in all frames of a given video. For video scene parsing task, pixel-by-pixel feature computation leads to high computational cost, which is particularly prominent in large-scale video scenes. For a given scene video, a certain degree of downsampling can greatly reduce the computational cost, but the accompanying cost is that the details of the video scene are ignored. How to effectively deal with the high computational cost and propose a real-time video scene parsing algorithm is one of the future research directions. Furthermore, how to combine local spatio-temporal information with global features to generate better feature representation is worth further research. In addition, indoor scenes often contain many man-made objects (beds, tables, chairs, etc.), which have a more linear structure. How to extract geometric structures to represent indoor objects is also worth exploring.
- Human Action Recognition. In the last ten years, the research on this subject has been quite good, and has achieved good research results. Generally speaking, the recognition of human movements is obtained through the form of video or image to identify, and then with the development of science and technology, as well as the development of the electronic industry, begin to use wearable sensor devices to identify human movements. With the development of wireless technology and

the expansion of coverage, WiFi signal has been used to identify human movements, and has achieved better results. At present, the subject of WiFi recognition of human movements is the latest research trend. The following processes are generally used to deal with human movement recognition. First of all, the data is collected, and the collected data is denoised or processed. Then the feature quantity is extracted, and the training and classification are carried out. Finally, the recognition of human movements is realized. Almost all research teams rely on this general process to recognize human movements. In these five parts, data denoising and feature extraction are two key links. Researchers are continuously improving and developing these two links in depth, improving the accuracy of recognition.

3D-based scene understanding task aims to understand the scene within a 3D environment, and the input information source can be the RGB-D data, LiDAR data, point-cloud data. Applications of 3D-based scene understanding are diverse, such as robotics, autonomous driving, augmented reality, virtual reality, and surveillance. It enables machines to perceive and interact with the world in a more human-like manner, which is crucial for many AI-driven applications.

- Point-cloud object segmentation aims to partition a 3D point cloud into meaningful segments or clusters. Here, each segment corresponding to a distinct object or part of the scene. This task is essential for various applications, such as robotics, autonomous driving, augmented reality, and object recognition. Incorporating data from multiple sensor modalities, such as LiDAR and RGB-D data, can provide richer information for segmentation. Fusion methods that effectively combine different types of data could lead to more accurate and robust segmentation results. Current segmentation approaches highly rely on large amounts of labeled data for training, which can be expensive and time-consuming to acquire. Future work may explore to design unsupervised learning or reinforcement learning network models to further improve the intelligence of point-cloud object segmentation.
- 3D Gesture Recognition. The application of gesture recognition is mostly applied to the control and perception of smart home, education and learning. In recent years, the rapid development of machine learning algorithms, particularly deep learning, may lead to improved accuracy and robustness in 3D gesture recognition task. Benefiting from this, some difficult gestures are also well recognized, such as viewpoint changes and occlusions. However, the real-time of 3D gesture recognition is essential for many applications, especially those involving interactive systems or virtual environments. Future work may focus on optimizing algorithms to achieve low-latency, real-time performance without reducing accuracy. At the same time, how to use multimodal information to assist 3D gesture recognition is also a future research direction.
- 3D Human Pose Estimation. ST-GCN is a spatiotemporal graph convolutional network that can be used for human behavior recognition. This algorithm models the dynamic skeleton based on the time series representation of human joint position, and extends the graph convolution to the spatiotemporal graph convolution network to capture this spatiotemporal variation. Skeleton-based data can

be obtained from motion capture devices, as well as pose estimation algorithms from video. Usually the data is a sequence of coordinate systems, each with a set of joint coordinates. ST-GCN is to construct a spatio-temporal graph which treats joints as graph nodes and considers the natural connections of human body as well as times as edges. The input to ST-GCN is the joint coordinate vector on the graph node. This can be seen as a simulation of an image-based cnn, where the input is made up of pixel intensity vectors residing on a 2D image grid. Multilevel convolution operation of spatiotemporal graph is carried out on input data to generate higher-level feature graph on the graph. It will then be sorted into the corresponding action category by the standard SoftMax classifier.

The manufacturer's authorised representative in the EU is Springer Nature Customer Service Centre GmbH, Europaplatz 3, 69115 Heidelberg, Germany. If you have any concerns regarding our products, please contact ProductSafety@springernature.com

Printed and bound by CPI Group (UK) Ltd, Croydon, CR0 4YY
01/10/2026
02225914-0005